COLLEGE MATHEMATICS

CLEP* Test Study Guide

© 2020 Breely Crush Publishing, LLC

CLEP is a registered trademark of the College Entrance Examination Board which does not endorse this book.

971010420143

Published by Breely Crush Publishing, LLC
10808 River Front Parkway
South Jordan, UT 84095
www.breelycrushpublishing.com

ISBN-10: 1-61433-632-6
ISBN-13: 978-1-61433-632-7

Printed and bound in the United States of America.

CLEP is a registered trademark of the College Entrance Examination Board which does not endorse this book.

Table of Contents

 Sets

What is a "set"? A set is a collection of elements. Elements can be any type of number. A finite set has a specific number of elements while an infinite set has an unlimited number of elements.

What are examples of sets? Elements of a set are placed within braces { } and each element is separated by a comma.

Example of a finite set: A = {3, 4, 5, 6, 7, 8}
This is a finite set because it has a specific number of elements. There are six elements in this set.

Example of an infinite set: W = {0, 1, 2, 3, 4, …}
The "…" denotes that the elements in the set continue on infinitely.
In this case, the set W represents all the Whole numbers.

What notation do you use with sets?
Set notation uses symbols for each definition. Following is a list of relationships between sets and their symbols.

Empty Set: An empty set is also referred to as a null set. An empty set does NOT have any elements in it and is represented by {} or $\varnothing$.

Common Mistake: It may be tempting to identify {0} as an empty set. However, {0} is not an empty set because it contains one element which is the number 0.

Union of Two Sets: A union of two sets means that you combine the elements of one set with the elements of the other set to create one, larger set. A union of two sets is denoted by $\cup$.

Example: Let's say you have two sets, set A and set B.
A = {1, 2, 3, 4} and B = {2, 4, 6, 8}.
A $\cup$ B = {1, 2, 3, 4, 6, 8}

Note: You don't write the numbers A and B have in common twice. For example, both A and B had "2" and "4" in common. You only write "2" and "4" ONCE in the union of sets A and B.

Intersection of Two Sets: An intersection of two sets means that you only write the elements that are common to both sets. An intersection of two sets is denoted by $\cap$.

Example: Let's say you have two sets, set C and set D.
C = {2, 3, 4, 5, 6, 7, 8} and D = {6, 7, 8, 9}.

C $\cap$ D = {6, 7, 8} because these are the only numbers that are in both set C and set D.

Note: What if the two sets have no elements in common? For example, what if you had A = {1, 3, 5} and B = {2, 4, 6, 8}? They don't have any numbers in common. In this case, the two sets are disjoint and you write their intersection as the empty set {}.

Subset: The $\subset$ symbol is used to denote a subset. A $\subset$ B means that A is a subset of B.

Example of a Subset: A = {1, 2, 3} and B = {1, 2, 3, 4, 5, 6}
Every number in set A is in set B therefore, A is a subset of B, A $\subset$ B.

However, B is NOT a subset of A. Why not? B $\not\subset$ A because set B has more elements than set A and therefore can't be a subset of set A.

Disjoint Sets: Disjoint sets are sets which have none of the same numbers. In other words, disjoint sets are sets which result in an empty set when intersected.

Example of disjoint sets:
A={2, 4, 6} and B={1, 3, 5}
Therefore A $\cap$ B={}

Equivalent Sets: Equivalent sets are sets which have all of the same numbers. In other words, if two sets are equivalent, then the intersection of the two sets is equal to each of the sets.

Example of equivalent sets:
A={1, 14, 21} and B={1, 14, 21}
Therefore A $\cap$ B={1, 14, 21}

🎓 *Venn Diagrams*

Are you a visual learner and find it easier to remember math concepts when using pictures or visual clues? Then Venn Diagrams are for you! Venn Diagrams represent the relationship between two or more sets visually.

Venn Diagram

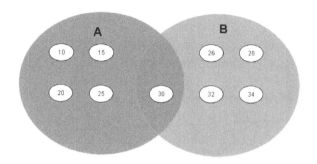

Each circle represents a set.
A = {10, 15, 20, 25, 30} B = {26, 28, 30, 32, 34}

A ∩ B = {30} which is represented by the overlap of the two circles.
A ∪ B = {10, 15, 20, 25, 26, 28, 30, 32, 34}

What is the Cartesian Product? Cartesian refers to the xy plane and product refers to multiplication. The Cartesian product involves finding the product of two sets of numbers. The following example demonstrates how to find the Cartesian product.

Let $A = \{1,2,3\}$ and $B = \{x,y,z\}$

The Cartesian product is written as $A \times B$
$A \times B = \{(1,x),(1,y),(1,z),(2,x),(2,y),(2,z),(3,x),(3,y),(3,z)\}$

"TRY IT YOURSELF" – QUESTIONS ABOUT SETS

Question 1: Given the following sets: {100, 101, 102}, {1, 2, 3, 4, …}, {…-3, -2, -1, 0, 1, 2, 3,…}, {44, 55, 66}, {1.5, 2.8, 10.4}, $\{\frac{1}{2},\frac{1}{4},\frac{1}{8},\frac{1}{16}\}$.

Which of these sets are finite sets?
Which of these sets are infinite sets?

Answer 1: Finite sets have a specific number of elements. The finite sets include {100, 101, 102}, {44, 55, 66}, {1.5, 2.8, 10.4}, $\{\frac{1}{2},\frac{1}{4},\frac{1}{8},\frac{1}{16}\}$.

Infinite sets continue on infinitely in the positive or negative direction. The infinite sets include {1, 2, 3, 4, …}, {…-3, -2, -1, 0, 1, 2, 3,…}.

****

Question 2: Let A = {1, 3, 5, 7} and B = {1, 2, 3, 4, 5, 6, 7, 8}. What is A $\bigcup$ B?

Answer 2: $\bigcup$ means the Union of two sets. A $\bigcup$ B = {1, 2, 3, 4, 5, 6, 7, 8}.
Note: If the two sets have numbers in common then only write those numbers once.

****

Question 3: The union of which two sets gives the set F = {2, 4, 7, 10, 15, 29}?
A = {15, 29}   B = {2, 4, 15}   C = {1, 2, 4, 7, 10}   D = {2, 7, 10, 29}

Answer 3: B $\bigcup$ D.
Note: Set A can't be included because it can't be combined with any of the other sets and have all the numbers present in set C. Set C can't be included because it includes the number 1, which is not present in set F.

****

Question 4: Let A = {5, 10, 15, 20} and B = {1, 10, 20, 30}. What is A $\bigcap$ B?

Answer 4: $\bigcap$ means the Intersection of two sets. A $\bigcap$ B = {10, 20}.
An intersection only includes the numbers that are common to both sets.

****

Question 5: The intersection of which two sets gives the set F = {25, 50, 75}?
A = {10, 15, 20, 25, 50, 55, 60, 75}   B = {25, 36, 47, 50, 65, 75, 86}   C = {25, 50, 65}
D = {50, 75, 125}

Answer 5: A $\bigcap$ B.
Note: Sets C does not include the number 75 and Set D does not include the number 25 so they can't be part of the intersection to form set F.

****

Question 6: Given the sets A = {5, 10, 15, 20, 25, 30, 35, 40} and B = {10, 20, 30, 40}.
Write True or False for each relationship and give your reasoning.
A $\subset$ B: _____
B $\subset$ A: _____

Answer 6: A ⊂ B = False, Reasoning: ⊂ means subset. A ⊂ B is read as "A is a subset of B" which means that every number in set A is in set B. This is false because set A contains more numbers then set B.

B ⊂ A = True, Reasoning: B ⊂ A means "B is a subset of A." Every number in set B is in set A. In addition, set B has fewer numbers then set A.

****

Question 7:

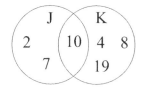

Fill in the blanks regarding the Venn Diagram:

J = _____   K = _____

J∩K = _____   J∪K = _____

Answer 7: J is the set of numbers in circle J. K is the set of numbers in circle K.
J = {2, 7, 10}   K = {4, 8, 10, 19}
J∩K = {10}   J∪K = {2, 4, 7, 8, 10, 19}

****

Question 8: Let A = {1, 10, 20} and B = {2, 40, 80, 100}. What is A×B?

Answer 8: A×B = {(1, 2), (1, 40), (1, 80), (1, 100), (10, 2), (10, 40), (10, 80), (10, 100), (20, 2), (20, 40), (20, 80), (20, 100)}
Note: × is the Cartesian product.

Logic

What is logic based upon? Logic is related to given *statements*. A statement can be true or false. When a statement is true, its truth value is T. When a statement is false, its truth value is F. For example, 1+4=5 has a truth value of T.

Statements may include connective statements including: "and", "or", "not", "if... then", and "...if and only if..." Each connective has a symbol and formal name associated with it. Study the following table regarding this information.

Connective	Formal Name	Symbol
and	Conjunction	$\wedge$
or	Disjunction	$\vee$
not	Negation	$\neg$
If...then	Conditional	$\rightarrow$
...if and only if...	Biconditional	$\leftrightarrow$

What are Truth Tables?
Elementary truth tables represent those truth values of two statements with a specific connective. Let's use Statement 1 = S1 and Statement 2 = S2 to create truth tables.

Examine the truth table below.

S1　S2	S1 $\vee$ S2	S1 $\wedge$ S2	S1 $\rightarrow$ S2	S1 $\leftrightarrow$ S2
T　T	T	T	T	T
T　F	T	F	F	F
F　T	T	F	T	F
F　F	F	F	T	T

How do you read a Truth Table? Notice the first column and second row has "T" for both S1 and S2. This means that Statement 1 and Statement 2 were found to be true. Read across the second row to see how the "connectives" affect the truth value of the statement.

What are conditional statements? Conditional statements are written as "If...then" statements which include a hypothesis and conclusion. The hypothesis is what is given and the conclusion is what is to be proved.

For example, the statement you are given is "If a triangle has a right angle then it is a right triangle." The hypothesis is "If a triangle has a right angle" and the conclusion is "then it is a right triangle."

What is a counterexample? A counterexample is a statement which proves a rule false by presenting an exception to it.

For example, if your rule was "all boys are tall", then a counterexample could be if Tom was a boy who was short. Because Tom was an exception to the rule, or counterexample, the rule was proven false.

What is the Converse of a statement? The converse of a statement is the statement that is formed by switching the hypothesis with the conclusion.

For example, if your statement was "cats are tigers" then the converse of the statement would be "tigers are cats."

What is the Negative of a statement? The negative of a statement is the denial of a statement.

For example, if your statement was "a boy is tall" then the negative of the statement would be "a boy is not tall."

What is the Inverse of a statement? The inverse of a statement is formed by denying both the hypothesis and the conclusion.

For example, if your statement was "a boy is tall" then the inverse statement would be "a person who is not a boy is not tall."

What is the Contrapositive of a statement? The Contrapositive of a statement is created by switching the negative of the hypothesis with the negative of the conclusion.

What are necessary and sufficient conditions? Necessary and sufficient conditions are utilized to ascertain if the conditions in the hypothesis are "necessary or sufficient" to verify its conclusion.

How do you determine if you have necessary and sufficient conditions?

1. First you need to determine whether you statement is true or false.
2. Second determine if the converse of the statement is true or false.
3. Apply the following four principles:

Principle 1: If a statement and its converse are both true, then the conditions in the hypothesis of the statement are necessary and sufficient for its conclusion.

Principle 2: If a statement is true and its converse is false, then the conditions in the hypothesis of the statement are sufficient but not necessary for its conclusion.

Principle 3: If a statement is false and its converse is true, then the conditions in the hypothesis are necessary but not sufficient for its conclusion.

Principle 4: If a statement and its converse are both false, then the conditions in the hypothesis are neither necessary nor sufficient for its conclusion.

"TRY IT YOURSELF" – QUESTIONS ABOUT LOGIC

Question 1: What is the converse of the statement "squares are rectangles"?

Answer 1: "rectangles are squares"
Explanation: The hypothesis of the original statement is "squares" and the conclusion is "rectangles." The converse of a statement switches the order of the hypothesis and conclusion.

****

Question 2: What is the negative of the statement "squares are rectangles"?

Answer 2: "squares are not rectangles"

****

Question 3: What is the Inverse of the statement "squares are rectangles"?

Answer 3: "shapes that are not squares are not rectangles"
Explanation: The Inverse of the statement denies both the hypothesis and the conclusion.

****

Question 4: What is the Contrapositive of the statement "squares are rectangles"?

Answer 4: "shapes that are not rectangles are not squares"
Explanation: The Contrapositive of a statement is when you switch the negative of the hypothesis with the negative of the conclusion.

****

Question 5: Determine whether the conditions in the hypothesis are necessary or sufficient to justify the conclusion for the statement, "squares are rectangles."

Answer 5: The conditions in the hypothesis of the statement are sufficient, but not necessary for its conclusion.

Explanation: When determining necessary and sufficient conditions answer these two questions:

1) Is the statement true or false? Squares are rectangles so the statement is true.

2) Is the converse of the statement true or false? The converse is false because rectangles are not squares.

Principle 2 says that if the statement is true and the converse is false, then the conditions in the hypothesis of the statement are sufficient, but not necessary for its conclusion.

****

Question 6: Determine whether the conditions in the hypothesis are necessary or sufficient to justify the conclusion for the statement, "rectangles are squares."

Answer 6: The conditions in the hypothesis are necessary, but not sufficient for its conclusion.

Explanation: The statement is false. The converse of the statement is true. This follows Principle 3 which says the conditions in the hypothesis are necessary, but not sufficient for its conclusion.

****

Question 7: What is the difference between the Negative of a statement and Inverse of a statement?

Answer 7: A negative of a statement is a denial of the hypothesis. An Inverse of a statement is a denial of the hypothesis and a denial of the conclusion.

****

Question 8: Is the following statement true or false? Explain your answer. "If a triangle has three different side lengths then it is a scalene triangle."

Answer 8: False. A scalene triangle does have three different side lengths, but a right triangle has three different side lengths as well.

****

Question 9: Give the formal names for each connective in the following statements.
"An angle is acute if and only if it is less than 90°" _____
"Triangles and squares are geometric shapes." _____
"If a number can be written as a fraction then it is a rational number." _____
"Squares are rectangles, but rectangles are not squares." _____
"The number 2 can be referred to as an Integer, Whole number, Natural number, or Rational number." _____

Answer 9:
"An angle is acute if and only if it is less than 90°": **biconditional**
"Triangles and squares are geometric shapes." **conjunction**
"If a number can be written as a fraction then it is a rational number." **conditional**
"Squares are rectangles, but rectangles are not squares." **negation**
"The number 2 can be referred to as an Integer, Whole number, Natural number, or Rational number." **disjunction**

Example of Determination of Necessary and Sufficient Conditions

For each of the following two statements, determine whether the conditions in the hypothesis are necessary or sufficient to justify the conclusion:

Statement #1: If a man lives in New York City then he lives in New York State.

Answer: Statement #1 is true.
The converse is "If a man lives in New York State then he lives in New York City." The converse is false. We follow Principle 2 and the hypothesis is sufficient, but not necessary.

Statement #2: An equilateral polygon is equiangular.

Answer: Statement #2 is false. The converse is "An equiangular polygon is equilateral" which is false. Therefore, by Principle 4 the conditions in the hypothesis are neither necessary nor sufficient.

 # Real Number System

What is the real number system? The real number system consists of many different types of numbers including:

- Natural numbers
- Whole numbers
- Integers
- Rational numbers
- Irrational numbers

Natural Numbers include {1, 2, 3, 4, 5,...}

Whole Numbers include {0, 1, 2, 3, 4, 5,...}

Integers include {..., -4, -3, -2, -1, 0, 1, 2, 3, 4,...}

Rational Numbers: Rational numbers are numbers that can be expressed as a fraction. The fraction can be proper or improper and positive or negative. Examples include:

$$\frac{7}{11}, -\frac{7}{11}, \frac{9}{8}, -\frac{9}{8}$$

Irrational Numbers: Numbers that can't be expressed as a fraction. Irrational numbers include numbers decimals that continue on forever such as π. If a number is not rational then it is irrational.

"TRY IT YOURSELF" – QUESTIONS ABOUT THE REAL NUMBER SYSTEM

Question 1: Given the set of numbers $\{-5, 0, 7, 3.4, \frac{2}{3}, \sqrt{36}, \sqrt{11}, \frac{-1}{4}\}$.

Which numbers are rational numbers? _____

Which numbers are whole numbers? _____

Which numbers are natural numbers? _____

Which numbers are integers? _____

Which numbers are irrational numbers? _____

Answer 1:

What numbers are rational numbers? $\{-5, 0, 7, 3.4, \frac{2}{3}, \sqrt{36}, \frac{-1}{4}\}$ *Rational numbers can be written as fractions.*

Which numbers are whole numbers? $\{0, 7, \sqrt{36}\}$ *Whole numbers are 0, 1, 2, 3...*

Which numbers are natural numbers? $\{7, \sqrt{36}\}$ *Natural numbers are 1, 2, 3...*

Which numbers are integers? $\{-5, 0, 7, \sqrt{36}\}$ *Integers can be positive, negative, and 0. They are ...-3, -2, -1, 0, 1, 2, 3...*

Which numbers are irrational numbers? $\sqrt{11}$ *Irrational numbers can't be expressed as a fraction or decimal that terminates or repeats.*

****

Question 2: Given the set of numbers $\{2, -1, 11, \frac{3}{7}, 4, 8, 12.5, 4\frac{1}{2}\}$.

Which numbers are even numbers? _____
Which numbers are odd numbers? _____
Which numbers are prime numbers? _____
Which numbers are composite numbers? _____

Answer 2:
Which numbers are even numbers? $\{2, 4, 8\}$
Which numbers are odd numbers? $\{-1, 11\}$
Which numbers are prime numbers? $\{2, 11\}$ *A prime number is a number greater than 1 that is only divisible by itself and 1.*
Which numbers are composite numbers? $\{4, 8\}$ *A composite number is a number that can be divided by other numbers besides 1 and itself.*

****

Question 3: What are the factors of the number 24?

Answer 3: The factors of 24 include: 1, 2, 3, 4, 6, 8, 12, and 24.

****

Question 4: What are the factors of the number 27?

Answer 4: 1, 3, 9, and 27.

****

Question 5: What is the greatest common factor between 25 and 40?

Answer 5: The greatest common factor is 5.
Explanation: The factors of 25 are 1, 5, and 25. The factors of 40 are 1, 2, 4, 5, 8, 10, 20, and 40. The greatest number they have in common is 5.

****

Question 6: What is the greatest common factor between 24 and 27?

Answer 6: 3
Explanation: The only factor the numbers 24 and 27 have in common is 3.

****

Question 7: What is the absolute value of -10?

Answer 7: 10. Absolute value of a number is the positive value of the number.

****

Question 8: What is -|-25|?

Answer 8: -25. The absolute value of -25 is 25, but there is an additional negative sign on the outside of the absolute value sign. This additional negative sign turns the 25 to -25.

****

Question 9: What inequality should be written in the blank? $\left|\dfrac{-2}{3}\right| \underline{\quad\quad} -\left|\dfrac{-2}{3}\right|$

Answer 9: > because it is $\dfrac{2}{3} > -\dfrac{2}{3}$.

****

Question 10: What is the order of operations? Should you work from left to right or right to left?

Answer 10: Parentheses, Exponents, Multiplication & Division, Addition & Subtraction. Work from left to right. (Remember: "PEMDAS.")

****

Question 11: Evaluate: $10 + (26 \div 2) + 100/5 - 4^2$

Answer 11: 27.
Explanation: Step 1: Evaluate inside the parenthesis first $(26 \div 2) = 13$.
Step 2: Evaluate the exponent. $4^2 = 16$.
Step 3: Division $100/5 = 20$
Step 4: Add and subtract from the left to right: $10 + 13 + 20 - 16 = 27$.

****

Question 12: Evaluate: $|6 + (5^2) - 48|$

Answer 12: 17.
Explanation: Step 1: Evaluate the exponent in the parenthesis. $(5^2) = 25$.
Step 2: Add and subtract from left to right. $|6 + 25 - 48| = |\text{-}17|$
Step 3: Find the absolute value of $|\text{-}17| = 17$.

****

Question 13: Evaluate: $500 \div 250 + [(12 \times 4) + 20] + 30 \times 10$.

Answer 13: 370.
Explanation: Step 1: Evaluate the innermost parenthesis. $(12 \times 4) = 48$.
Step 2: Evaluate the outermost parenthesis. $[48 + 20] = 68$.
Step 3: Multiply and divide from left to right. $500 \div 250 = 2$ and $30 \times 10 = 300$.
Step 4: Add and subtract from left to right. $2 + 68 + 300 = 370$.

****

Question 14: Evaluate: $((\frac{1}{2})^2 - (44 \div 11)) + \frac{3}{4}$

Answer 14: -3.

Explanation: Step 1: Evaluate the innermost parentheses. $((\frac{1}{2})^2) = \frac{1}{4}$ and $(44 \div 11) = 4$.

Step 2: Evaluate the outermost parenthesis. $(\frac{1}{4} - 4) = \frac{-15}{4}$.

Step 3: Add and subtract from left to right. Simplify. $\frac{-15}{4} + \frac{3}{4} = \frac{-12}{4} = -3$

****

Question 15: Given a set of numbers: {600, 111, 48, 955, 711, 118}
Answer each question and give the rule of divisibility that was used.

Which numbers are divisible by 2? _____
Which numbers are divisible by 5? _____
Which numbers are divisible by 10? _____
Which numbers are divisible by 3? _____
Which numbers are divisible by 9? _____

Answer 15:
Which numbers are divisible by 2? 600, 48, 118.
Rule: Numbers divisible by 2 end in 0, 2, 4, 6, or 8.
Which numbers are divisible by 5? 600, 955.
Rule: Numbers divisible by 5 end in either 0 or 5.
Which numbers are divisible by 10? 600.
Rule: Numbers divisible by 10 end in 0.
Which numbers are divisible by 3? 600, 111, 48, 711.
Rule: If the digits of the number add up to a multiple of 3 then the number is divisible by 3.
Which numbers are divisible by 9? 711.
Rule: If the digits of the number add up to 9 then the number is divisible by 9.

****

Question 16: Given the set of numbers: $\{\pi, 2.\overline{33}, \sqrt{100}, -2.5, -4\}$

Which numbers are rational numbers? _____
Which numbers are irrational numbers? _____
Which numbers are whole numbers? _____
Which numbers are natural numbers? _____
Which numbers are integers? _____

Answer 16:

Which numbers are rational numbers? $2.\overline{33}, \sqrt{100}, -2.5, -4$ *Note: $2.\overline{33}$ is $2\frac{1}{3}$*

Which numbers are irrational numbers? π

Which numbers are whole numbers? $\sqrt{100}$

Which numbers are natural numbers? $\sqrt{100}$

Which numbers are integers? $\sqrt{100}$, -4

Basic Properties of Numbers

What is an Even number? An even number is any number that is divisible by 2 and gives you an integer for your answer. An integer type answer means that you will not have a decimal or fractional value in your answer.

Examples of Even Numbers: 2, 4, 6, 8

What is an Odd number? An odd number is any number that when divided by 2 will give you a decimal or fractional value in your answer.

Examples of Odd Numbers: 1, 3, 5, 7

What is a prime number? A prime number is a number greater than 1 that is only divisible by itself and 1. When we say "divisible" we mean that the answer will come out to be an integer and NOT have a decimal or fractional value.

Examples of Prime Numbers: 2, 3, 5, 7, 11, 13, 17,...
For example, 2 is a prime number because you can only divide 1 and 2 into the number 2 and come out with an integer answer.

Common Mistake: Remember "1" is NOT a prime number.

What is a composite number? A composite number is any number that is NOT prime. Or in other words, a composite number can be divided by other numbers besides 1 and itself.

Examples of Composite Numbers: 4, 6, 8, 10,…
For example, 6 is a composite number because it is divisible by 1, 2, 3, and 6.

Factors and Divisibility

What are factors? Factors are numbers that when multiplied together equal a specific number. For example, the factors of the number 6 are 1, 2, 3, and 6. Why? We know that $1 \times 6 = 6$ and $2 \times 3 = 6$ and this satisfies the definition of factors.

Math Terminology Note: The mathematical term "Product" means multiplication. Hence, the product of two factors should equal a specific number.

How do you figure out the factors of a number? You can try to figure out the factors of a number by trial and error. For example, if you have the number 10, you can try to think of all the numbers that when multiplied together equal 10. The factors of 10 are 1, 2, 5, and 10.

How do you find the Greatest Common Factor between two numbers? The Greatest Common Factor is the largest factor that the two numbers have in common.

Greatest Common Factor Method #1:
One method to find the Greatest Common Factor between two numbers it to make a list of the factors of each number and find the largest number they have in common.

For example, find the Greatest Common Factor between 48 and 60.

Factors of 48:
1, 2, 3, 4, 6, 8, 12, 16, 24, 48

Factors of 60:
1, 2, 3, 4, 5, 6, 10, 12, 15, 20, 30, 60

Therefore, the Greatest Common Factor or "GCF" between 48 and 60 is 12.

Greatest Common Factor Method #2:
Another method you can use to find the Greatest Common Factor between two numbers is the "Product of Primes" method. In this method you construct a tree diagram with your initial numbers at the top of their own trees. Draw branches that represent two numbers that when multiplied equal the number above them. Continue through this process until you are left with only Prime numbers at the bottom of the tree diagram.

Let's find the GCF between 48 and 60 using this method. Following is a visual representation of the "Product of Primes" for clarification.

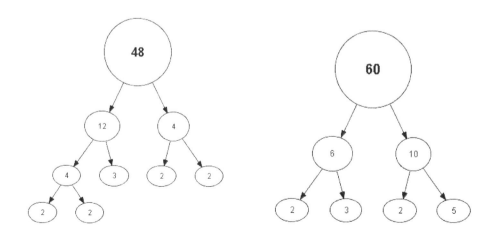

It doesn't matter what numbers you multiply throughout the tree as you will eventually come to the same prime numbers at the bottom of the tree.

To find the GCF find the numbers that each tree has in common.

The Product of Primes for 48 was $2 \times 2 \times 2 \times 2 \times 3$
The Product of Primes for 60 was $2 \times 2 \times 3 \times 5$
The numbers in common are $2 \times 2 \times 3$. Multiply this out $2 \times 2 \times 3 = 12$.
The Greatest Common Factor between 48 and 60 is 12.

What is Divisibility? Divisibility refers to the fact that when you find the factors of a number you can also consider them "divisors" of that number. For example, we know that 1, 2, 3, and 6 are factors of 6. This means that 1, 2, 3, and 6 are also divisors of 6.

What are rules of Divisibility? There are rules you can follow when dividing that are considered shortcuts to the final answer.

Common Divisibility Rules:

1. Numbers divisible by 2 end in 0, 2, 4, 6, or 8.

2. Numbers divisible by 5 end in either 0 or 5.

3. Numbers divisible by 10 end in 0.

4. If the digits of the number add up to 3 then the number is divisible by 3.
For example, 336 is divisible by 3 because $3 + 3 + 6 = 12$

5. If the digits of the number add up to 9 then the number is divisible by 9.
For example, 45 is divisible by 9 because $4 + 5 = 9$

 # *Absolute Value and Order*

What is absolute value and why do we need it? Absolute value refers to the distance between a number and 0 on the number line. Absolute value is always positive. We need it because we can't have a negative distance just as we can't have negative time. It doesn't make sense because we can't go back in time.

What is the notation for Absolute Value and how do we find it? The notation for absolute value is $|\ |$. You change the sign of the number inside the $|\ |$ to a positive number.

Examples of Absolute Value:

$$|-2| = 2 \qquad |4| = 4 \qquad \left|\frac{-3}{4}\right| = \frac{3}{4}$$

Note: $-|-2| = -2$ Remember you have two negative signs in this problem. The negative sign inside of the Absolute Value changes to positive, but the negative sign on the outside of the Absolute Value remains negative.

What does order have to do with Absolute Value? Order refers to absolute values that come in this form: $|5-8|$. When you have more than one number inside of the Absolute Value then you can think of the Absolute Value bars as parentheses and use the order of operations evaluate the numbers.

Remember Order of Operations is Parentheses, Exponents, Multiplication & Division, and Addition & Subtraction. Work from left to right in each problem.

Examples of Absolute Value and Order of Operations:

$$|5-8| = |-3| = 3 \qquad |1 \times 3 + (5-3) - 2^3| = |-3| = 3$$

Open and Closed Intervals

What are opened and closed intervals? The real number system can be represented on a number line. There is a real number that matches each point on the number line.

Number Line: $\quad\quad\quad\quad -5 \quad\quad\quad 0 \quad\quad\quad 5$
$\quad\quad\quad\quad\quad\quad\quad\quad\quad\downarrow\quad\quad\quad\downarrow\quad\quad\quad\downarrow$

A **closed interval** includes endpoints, and is denoted by shaded circles.

An **open interval** does NOT include the endpoints, and is denoted by open, or un-shaded, circles.

Functions and Their Graphs

What is a function? A mathematical function represents a systematic manner in which to find a *value*. In other words, when you input information into a function you will generate a specific "unique" output. A function is usually written as $f(x)$.

For example, let's look at the function: $f(x) = x + 2$ and the value of $x = 3$

The Input is the value $x = 3$ while the output will be what you find when you plug "3" in for x in the equation $f(x) = x + 2$.

$$f(x) = x + 2$$
$$f(3) = 3 + 2$$
$$\text{Therefore, } f(3) = 5$$

Can you figure out what the value of $f(10)$ given $f(x) = x + 2$?

$$f(x) = x + 2$$
$$f(10) = 10 + 2$$
$$\text{Therefore, } f(10) = 12$$

Visual example:

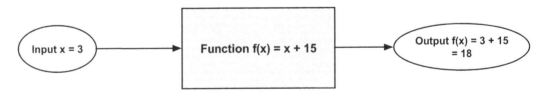

How do functions relate to graphing? Functions pair up x values with y values. Functions must have only one **y** (or output) value for each **x** (or input). Following is a visual representation to demonstrate this point.

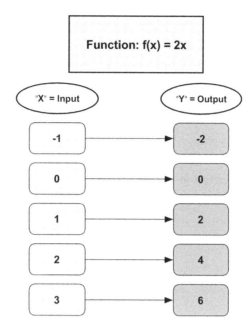

You can think of each "x" and "y" value as an **ordered pair (x,y)** in a graph. The ordered pairs in the previous example include: (-1, -2), (0,0), (1,2), (2,4), (3,6).

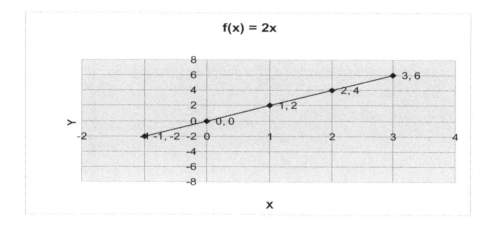

What is the domain of a function? The domain of a function refers to all the possible values you can use for "x" in the function $f(x)$ otherwise known as the first number in each ordered pair. The function $f(x) = 2x$ produced the following ordered pairs: (-1, -2), (0,0), (1,2), (2,4), (3,6). The Domain represents the highlighted numbers which are {-1, 0, 1, 2, and 3}.

What is the range of a function? The range of a function refers to all the possible values you can determine for the output $f(x)$ or "y." The range from the previous $f(x) = 2x$ is the second number in each ordered pair. (-1, -2), (0, 0), (1, 2), (2, 4), (3, 6). Therefore, the range is {-2, 0, 2, 4, 6}

How can you tell if a graph represents a function? Use the "vertical line test" to determine if a graph represents a function. The vertical line test states that if you draw a vertical line through a piece of the graph and it intersects the graph at one more than one point then it is NOT a function. (Remember for every "x" value you can have only ONE "y" value in order for the relationship to be a function.)

Following is a visual representation to explain the vertical line test.

$f(x) = x^2$ is a function as the red vertical line passes through the graph at only one y value.

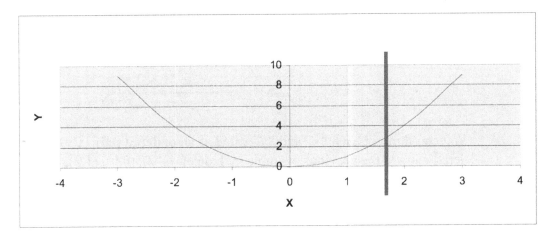

The following graph is NOT a function because the red vertical line passes through more then one y value.

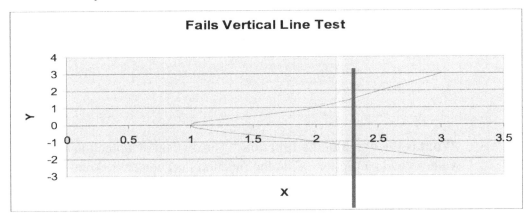

How do you find the Composition of two functions? The composition of two functions takes the terms of one function and substitutes them into the other function. The composition of two functions is denoted as $f(g(x))$.

For example, find $f(g(x))$

When $f(x) = x^2 + 12$ and $g(x) = 3x$

$f(g(x)) = f(3x) = (3x)^2 + 12 = 9x^2 + 12$

Therefore, $f(g(x)) = 9x^2 + 12$

How do you find the Inverse of a function? The inverse of a function can be found by switching the x variable with the y variable. For example, let $f(x) = 3x + 5$ then the inverse is denoted as f^{-1} is $f^{-1}(x) = \dfrac{x-5}{3}$.

 # Simple Transformations of Functions

What is Symmetry? Many graphs can be described as symmetric. A graph can be symmetric about the y axis, the x axis, and the origin. A graph which is symmetric about the y axis will have the same y coordinate when evaluating f(x) as when evaluating f(-x) (See example a). A graph which is symmetric about the x axis would have points appearing at both (x, y) and (x, -y) (see example b). Therefore are graph which is symmetric about the x axis is not a function. A graph which is symmetric about the origin will have opposite values of y coordinates when evaluating f(x) and f(-x) (see example c).

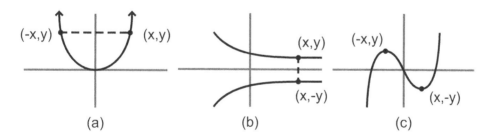

(a)   (b)   (c)

What are Translations? A translation is when the shape of the function is kept the same, but it is shifted to a different place on the coordinate graph.

A vertical translation is when the graph shifts up or down on the y axis. Vertical translations occur when a number is added to the end of the function.

For example, the graphs of $y = x^2$ and $y = x^2 + 3$ here

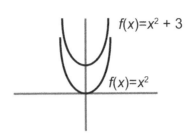

A horizontal translation is when the graph shifts right or left along the x axis. Horizontal translations occur when a number is added after the x, but still inside the square.

For example, the graphs of $y = x^2$ and $y = (x+3)^2$

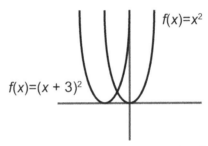

What are Reflections? A reflection is when the graph of a function is mirrored onto another part of the graph. Generally reflections occur across either the x or y axis. A graph will be symmetric across the line it is reflected over.

The graph of $y = -f(x)$ is the same as the graph of $y = f(x)$ reflected across the x axis. For example, $y = x^2$ and $y = -x^2$

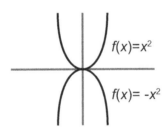

"TRY IT YOURSELF" – QUESTIONS ABOUT FUNCTIONS AND THEIR GRAPHS.

Question 1: Is y a function of x in $y = 15x$? Explain your reasoning.

Answer 1: Yes, y is a function of x in $y = 15x$ because for every x value there is a unique y value. This is the definition of a function. For example, if x = 1 then y = 15. If x = 2 then y = 30.

****

Question 2: Is y a function of x in $y = -\frac{2}{3}x$? Explain your reasoning.

Answer 2: Yes, y is a function of x in $y = -\frac{2}{3}x$ because for every x value there is a unique y value. For example, if x = 1 then $y = -\frac{2}{3}$ and if x = -1 then $y = \frac{2}{3}$

****

Question 3: Is y a function of x in $y = \sqrt{x}$? Explain your reasoning.

Answer 3: No, y is not a function of x in $y = \sqrt{x}$ because if x is a positive number then y has two possible values. For example, if x = 4 then y = 2 or -2. If x = 16 then y = 4 or -4. The two solutions come from the property of square roots.

****

Question 4: Does the following graph represent a function? Explain what test you can use and your answer.

Answer 4: Yes, the graph represents a function. There is a unique y value for each x value. I used the vertical line test.

****

Question 5: Evaluate $f(x) = 10x + 15$; $f(2)$ and $f(-4)$

Answer 5: $f(2) = 35$ and $f(-4) = -25$
Explanation: $f(2) = 10(2) + 15 = 20 + 15 = 35$
$f(-4) = 10(-4) + 15 = -40 + 15 = -25$

****

Question 6: Evaluate $f(x) = x^3 + 20 - x^2 + 10 - x$; $f(2)$ and $f(-2)$

Answer 6: $f(2) = 32$ and $f(-2) = 20$
Explanation: $f(2) = 2^3 + 20 - 2^2 + 10 - 2 = 8 + 20 - 4 + 10 - 2 = 32$
$f(-2) = (-2)^3 + 20 - (-2)^2 + 10 - (-2) = -8 + 20 - 4 + 10 + 2 = 20$

****

Question 7: Evaluate $y = \dfrac{1}{x}$; $x = 2$ and $x = 3$

Answer 7: $y(2) = \dfrac{1}{2}$ and $y(3) = \dfrac{1}{3}$

****

Question 8: Ryan has a part time job and earns \$5.25 per hour. Write a function that represents his weekly income, I, from his part time job. Use h to represent the number of hours he works each week.
Also find the amount of money Ryan earns in a week if he works 20 hours per week.

Answer 8: I = 5.25h; Ryan earns \$105 each 20 hour work week.
Explanation: Income = (Income per hour)(Number of hours worked)
          I = (\$5.25)(h)
For 20 hours of work, I = (\$5.25/hr)(20 hrs) = \$105.

****

Question 9: A Plumber costs \$50/hour plus an initial fee of \$100 just for making the trip to your house. Write a function that represents the total cost, C, of hiring this Plumber. Use h to represent hours.
Also find the total cost, C, if the Plumber worked for 4 hours.

Answer 9: C = 100 + 50h; The cost of the Plumber for 4 hours work = \$300.
Explanation: The Plumber charges \$100 initial fee no matter how many hours he works. Then he earns \$50 per hour which is dependent upon the number of hours he works.
C = Initial fee + (\$50/hr)(Number of hours worked) = \$100 + (\$50/hr)(4 hrs) = \$100 + \$200 = \$300.

****

Question 10: Evaluate $f(g(x))$ when $f(x) = x^2$ and $g(x) = 4$

Answer 10: $f(g(x)) = 16$
Explanation: $f(g(x)) = f(4) = 4^2 = 16$

****

Question 11: Evaluate $f(g(x))$ when $f(x) = 20x - 6$ and $g(x) = x^2 + 2$

Answer 11: $f(g(x)) = 20x^2 + 34$
Explanation: $f(g(x)) = f(x^2 + 2) = 20(x^2 + 2) - 6$
$f(g(x)) = 20x^2 + 40 - 6$
$f(g(x)) = 20x^2 + 34$

****

Question 12: Evaluate $f(g(x))$ when $f(x) = \dfrac{1}{x}$ and $g(x) = \dfrac{1}{x}$

Answer 12: $f(g(x)) = x$

Explanation: $f(g(x)) = \dfrac{1}{\frac{1}{x}} = x$

****

Question 13: Given the following function table:

X	Y
1	6
2	12
3	18
4	24

Answer the question and provide an explanation.

What is the function? _____
What are the ordered pairs of the function? _____
What is the domain of the function? _____
What is the range of the function? _____

Answer 13:
What is the function? Y = 6X; *This can be derived from the x and y values. Each Y value is 6 times each X value.*

What are the ordered pairs of the function? (1, 6), (2, 12), (3, 18), (4, 24); *The ordered pairs are the x and y values given in the table written as (x, y).*

What is the domain of the function? Domain = {1, 2, 3, 4}. *The domain of a function includes the x-values in a function. From the table, the x values are 1, 2, 3, and 4.*

What is the range of the function? Range = {6, 12, 18, 24}. *The range of a function includes the y-values in a function. From the table, the y values are 6, 12, 18, and 24.*

****

Question 14: The circumference of a circle is given by the formula $C = 2\pi r$. Express the circumference in terms of its radius r.

Answer 14: $r = \dfrac{C}{2\pi}$

Explanation: To express the formula in terms of r you need to get r by itself on one side. This can be accomplished by dividing 2π into both sides of the formula.

****

Question 15: The Pythagorean Theorem for right triangles is $a^2 + b^2 = c^2$. Express the formula in terms of a^2.

Answer 15: $a^2 = c^2 - b^2$

****

Question 16: What is the Inverse of the function $y = \dfrac{1}{3}x$? Explain your answer.

Answer 16: $f^{-1}(x) = f(y) = 3x$
Explanation: The inverse of a function is found by switching the x variable with the y variable.

Probability and Statistics

What is probability? Probability is defined as the likelihood that an event will occur expressed as the ratio of the number of favorable outcomes in the set of outcomes divided by the total number of possible outcomes.

What are outcomes and events? An outcome is the result of an experiment or other situation involving uncertainty. An event is a collection of outcomes.

For example, let's say you toss a coin. What is the probability that the outcome will be a "Head"? There are two possible outcomes which are "Head" or "Tail." Therefore, the probability is ½.

What are the basic rules of probability? Probability is denoted with **P("event")**. Probability is written in terms of the probability that a specific "event" will occur. For example, to determine the probability that event A will occur you denote this as **P(A).**

NOTE: The value of P **must** be between 0 and 1. Therefore, $0 \leq P(A) \leq 1$.

This means that the probability that event A will NOT occur is denoted as P(Not A) = 1- P(A).

How do you find the probability of a simple event? A simple event consists of finding the probability of one event "A." For example, event "A" can be rolling a die once and having an outcome of the number 3.

What are Independent Events? Two events are independent if the outcome of one event does NOT affect the outcome of the other event.
Find the probability of rolling a "3" when you throw one die.

$$P(3) = \frac{\text{number of 3's on die}}{\text{Total number of numbers on die}} = \frac{1}{6}$$

Conversely, find the probability that you don't roll a "3" when you throw one die.

You can compute:

$$P(\text{Not } 3) = \frac{\text{Total number of numbers that aren't 3 on die}}{\text{Total number of numbers on die}} = \frac{5}{6}$$

However, you can compute this by following the probability rule that
P(Not 3) = 1 – P(3)

Therefore, $P(\text{Not } 3) = 1 - \frac{1}{6} = \frac{5}{6}$

What is Conditional Probability? Conditional probability looks at the probability that more than one event occurs. Conditional probability deals with dependent events and is also referred to as compound probability.

Conditional Probability is denoted as P(B|A) which says the "Probability of event B given that event A has occurred."

What are Dependent Events? Two events are dependent if the outcome of one event depends on the outcome of the other event.

When you have dependent events you need to use the following **multiplication** rule to find the conditional probability:

$$P(A \cap B) = P(B|A)P(A)$$

What are Mutually Exclusive Events? Mutually exclusive events are two events that can NOT happen at the same time.

Example of Dependent (or Compound) Probability:
You have a bag filled with 6 marbles. 2 marbles are White, 2 marbles are Green, and 2 marbles are Yellow. Note: The marbles will not be replaced or placed back in the bag.

What is the probability that you will first pick a Green marble and then pick a White marble without looking?

Total Number of Outcomes = 6 marbles

First Pick (6 marbles):

$$P(White) = \frac{2}{6}$$

$$P(Green) = \frac{2}{6}$$

$$P(Yellow) = \frac{2}{6}$$

Second Pick (5 marbles left in the bag):

$$P(White) = \frac{2}{5}$$

$$P(Green) = \frac{1}{5}$$

$$P(Yellow) = \frac{2}{5}$$

Therefore, P(Green ∩ White) = P(White|Green)P(Green)

$$P(\text{Green}) = \frac{2}{6} \qquad\qquad P(\text{White|Green}) = \frac{2}{5}$$

$$P(\text{Green} \cap \text{White}) = \frac{2}{5} * \frac{2}{6} = \frac{4}{30} = \frac{2}{15}$$

What are Permutations? A permutation is an ordered sequence of items taken from a set of distinct items <u>without replacement</u>.

To find the number of permutations of

$$_nP_r = \frac{n!}{(n-r)!}$$

Remember "!" means factorial where n! = n(n-1)(n-2)...(1)
3! = 3 * 2 * 1 = 6
6! = 6 * 5 * 4 * 3 * 2 * 1 = 720

Example of a Permutation: Find the number of permutations of the numbers 2, 4, 6 taken two at a time.

Therefore, n = number of distinct items = 3
r = number of numbers in each group = 2

$$_3P_2 = \frac{3!}{(3-2)!} \;\; = \;\; \frac{3!}{(1)!} \;\; = \;\; \frac{3*2*1}{1} \;\; = \; \mathbf{6}$$

You can also find the number of Permutations by listing all the possibilities:

2, 4
2, 6
4, 2   *Note: In Permutations that 2,6 and 6,2 are considered to be separate results.*
4, 6
6, 2
6, 4

What are Combinations? A combination is an unordered subset of items taken from a group of distinct items.

$$\binom{n}{k} = \frac{n!}{k!(n-k)!}$$

Example of a Combination: Find the number of combinations for the numbers 2, 4, 6 taken two at a time.

Therefore, n = number of distinct items = 3
k = number of numbers in each group = 2

$$\binom{n}{k} = \frac{n!}{k!(n-k)!} = \frac{3}{2} = \frac{3!}{2!(3-2)!} = \frac{3!}{2!(1)!} = \frac{3*2*1}{2*1*1} = \frac{6}{2} = 3$$

There are only 3 combinations which include:
2,4
2,6
4,6

Remember in Combinations 2,4 and 4,2 are considered the same result and are NOT counted twice.

How do you find the "Mean" from sample data? The "Mean" is also referred to as the "Arithmetic Mean."

The Arithmetic Mean is calculated by finding the **sum** of all "X" values and dividing by N. This is the general formula to calculate the arithmetic mean: $\dfrac{\sum\limits_{i=1}^{n} X_i}{N}$

Example of Arithmetic Mean:

What is the arithmetic mean of the numbers 10, 8, 6, 12, and 9?

$$\frac{\sum\limits_{i=1}^{n} X_i}{N} = (10 + 8 + 6 + 12 + 9) = 45 \text{ and } N = 5 \text{ because there 5 numbers in the}$$

list.

$$\text{Therefore, } \frac{\sum\limits_{i=1}^{n} X_i}{N} = \frac{45}{5} = 9$$

What is the "Mode"? The Mode is the number that appears the most number of times in the sample data.

For example, the Mode for the following set of data {10, 12, 14, 10, 8, 10, 7} is **10** because 10 appears the most number of times in the data set.

What is the "Median"? The Median is the middle value or the arithmetic mean of the middle values. The following example will demonstrate how to find the Median.

Sample Data = {3, 3, 5, 6, 7, 9, 10, 12, 14}

The Median = 7 because it has four numbers to the left of it and four numbers to the right of it in the sample data.

> NOTE: The data MUST be in numerical order to find the Median. For example, if you are given a sample data set of {5, 6, 12, 3, 3, 10, 7, 14, 9} you must first put it in numerical order: {3, 3, 5, 6, 7, 9, 10, 12, 14} to find the correct Median of 7.

Find the Median for the following set of numbers: {2, 3, 4, 5, 6, 7}

Since there is an even number of numbers in the data set you will need to find the arithmetic mean of the two middle numbers 4 & 5. (4+5)/2 = 4.5. Therefore, the Median is 4.5.

What is the "Range"? The Range is simply the difference between the largest data value and the smallest data value. For example, the Range in the set of {2, 4, 7, 8, 10, 12, 24} is 24-2 = **22.**

What is "Standard Deviation"? Standard Deviation relates data points to the Mean of the sample data set. In other words, standard deviation is the mean of how far away the data points are from the actual mean. When s is large, the data is spread out. When s is small, the data is closer together.

Standard Deviation (s) is calculated using: $s = \sqrt{\dfrac{\sum_{i=1}^{n}(X_i - \bar{X})^2}{N}}$

Where $\overline{X}$ = the Arithmetic Mean of the data set

X_i = the value of each number in the data set

N = Number of values in the data set

Example of Standard Deviation: Find the Standard Deviation of the following set of numbers: $\{12, 6, 7, 3, 15, 10, 18, 5\}$.

$$\overline{X} = \frac{12+6+7+3+15+10+18+5}{8} = \frac{76}{8} = 9.5$$

$$s = \sqrt{\frac{(12-9.5)^2+(6-9.5)^2+(7-9.5)^2+(3-9.5)^2+(15-9.5)^2+(10-9.5)^2+(18-9.5)^2+(5-9.5)^2}{8}}$$

Therefore, $s = \sqrt{23.75} = 4.87$

What are the properties of Standard Deviation? The properties of Standard Deviation include:

1. $$s = \sqrt{\sum_{i=1}^{n}(X_i - \overline{X}}$$

2. You will often study "Normal Distribution." A Normal Distribution can be broken down into standard deviation from the mean. Following is the normal distribution curve.

The center of the Normal Distribution represents the Mean value. You can go 1 standard deviation in either the left or right direction by calculating $\overline{X}$ - s or $\overline{X}$ + s.

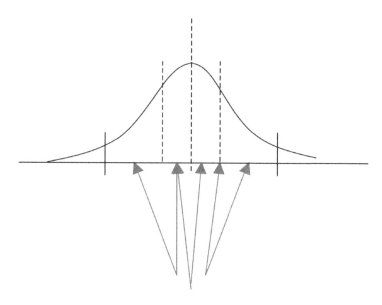

For a Normal Distribution:

- - - 68.27% of the numbers are within one standard deviation of the mean.

----- 95.45% of the numbers are within two standard deviations of the mean.

—— 98% of the numbers are within three standard deviations of the mean.

Data Interpretation and Representation

What are Quantitative and Qualitative Variables? There are two basic types of variables. Quantitative variables are numerical data. Temperature, time, and age are all quantitative variables because they can be represented by a number. Qualitative variables are variables that are represented by a category or label. Gender, political party, and hometown are all examples of qualitative variables.

Tables: Tables are generally the simplest way to organize numerical data, or two sets of quantitative variables. In math for example, x and y values are often organized into a table with one column representing x values, and a second column representing the corresponding y values.

The following table is an example of a table representing the values for $f(x) = x^2$ here.

X	Y
1	1
2	4
3	9
4	16

Scatterplots: Scatterplots are used to determine the relationship between two quantitative variables. The two variables are defined with one as dependent or explanatory, and the other as independent or response. The independent variable is plotted along the x axis, and the dependent variable is plotted along the y axis.

The following is an example of a scatterplot. The scatterplot shows the relation between height and age for a random group of students in an elementary school. Both of the variables (age and height) are quantitative, and each data point (meaning each set of age and height) are plotted appropriately.

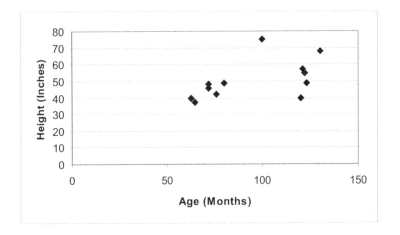

Line Graphs: Like scatterplots, line graphs display the relation between two quantitative variables. Line graphs are the best type of graph to use to show a change in one variable over time (the second variable).

The following example shows a line graph representing the average price of gasoline per gallon in the U.S. The independent variable (time) is plotted against the dependent variable (price per gallon), and the relationship is shown by the graph.

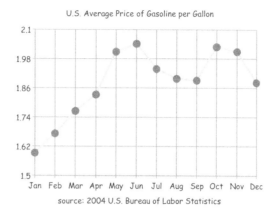

U.S. Average Price of Gasoline per Gallon

source: 2004 U.S. Bureau of Labor Statistics

Pie Charts: Pie charts are used when showing the relationship between categorical, or qualitative values. A pie chart is generally used when you want to show how one group of data relates to the whole. For example, if there are 21 boys and 15 girls in a class, the resulting pie chart would have two sections. The section representing boys will be larger, because proportional to the whole (the class) there are more boys.

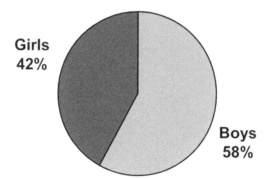

Histograms: Histograms are used with a single quantitative value. Mainly, histograms help show the distribution of data. The variable is listed along the x axis, however instead of using individual values, histograms use ranges. Along the y axis you use the number of data points that fall within the range, determining the height of the bar. Because of this, adjacent columns are placed next to each other with the sides touching. For example, the following histogram shows the distribution of ages of United States presidents at the time of their inauguration.

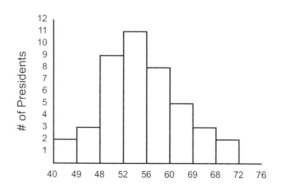

Bar Graphs: Bar graphs are very similar to histograms. However they are used with a single qualitative variable, instead of a quantitative one. Bar graphs are also used to show distribution. The independent variable is listed along the x axis, and will be a category. The y axis, like with histograms, shows the number of data points that fit each respective category.

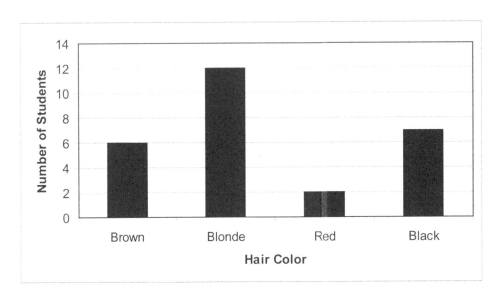

"TRY IT YOURSELF" – QUESTIONS ABOUT PROBABILITY.

Question 1: What is the probability of rolling a "4" on a six-sided number cube? Explain your answer.

Answer 1: P(4) = 1/6. *A six-sided number cube has 6 possible outcomes. The number "4" occurs once on the number cube. Therefore, P (4) = 1/6.*

****

Question 2: What is the probability of rolling an even number on a six-sided number cube? Explain your answer.

Answer 2: P (Even number) = 3/6 = 1/2. *There are 3 even numbers {2, 4, and 6} on a six-sided number cube. Therefore, P (Even number) = 3/6 = 1/2.*

****

Question 3: What is the probability of rolling a number greater then "1" on a six-sided number cube? Explain your answer.

Answer 3: P (Number > 1) = 5/6. *There are 5 numbers greater than 1 {2, 3, 4, 5, and 6} on the number cube. Therefore, P (Number > 1) = 5/6.*

****

Question 4: What is the probability of NOT rolling a number greater then "1" on a six-sided number cube? Explain two methods you can use to find this answer.

Answer 4: P (Not a number > 1) = 1/6.
Method 1: Use the complement: P (Not a number > 1) = 1 – P (Number > 1). P (Not a number > 1) = 1 – 5/6 = 1/6.

Method 2: There is one number less than 1 {1} on the number cube. Therefore, P (Not a number > 1) = 1/6.

****

Question 5: "A" is an event. Explain why P(A) = 1 – P(Not A) and P(Not A) = 1 – P(A).

Answer 5: P(A) is the probability that event "A" occurs. P(Not A) is the probability that all the outcome events, EXCEPT event "A" will occur. Probability must be a number between 0 and 1. Therefore, P(A) = 1 – P(Not A) and P(Not A) = 1 – P(A).

****

Question 6: Let Event A = rolling a number cube and Event B = flipping a fair coin. Are Event A and Event B independent or dependent events? Explain your answer.

Answer 6: Event A and Event B are independent events. They are independent events because the outcome of Event A does NOT affect the outcome of Event B.

****

Question 7: A card is drawn at random from a hat containing 14 black cards, 3 pink cards, 9 purple cards, and 18 yellow cards.
What is the probability of drawing a pink card? _____
What is the probability of drawing a black card or a purple card? _____
What is the probability of drawing a green card? _____
What is the probability of not drawing a yellow card? _____

Answer 7:
What is the probability of drawing a pink card? **P (Pink card) = 3/44**
What is the probability of drawing a black card or a purple card? **P (Black or Purple card) = 23/44**
What is the probability of drawing a green card? **P(Green card) = 0; There are no green cards in the hat.**
What is the probability of not drawing a yellow card? **P(Not a Yellow card) = 26/44 = 13/22**

****

Question 8: You have a box full of 10 shirts. 3 shirts are blue, 1 shirt is red, and 6 shirts are white.
Let Event A = Pick one shirt from the box without looking and you do not put the shirt back in the box.
Let Event B = Pick a second shirt from the box without looking.

Are Event A and Event B independent or dependent? Explain your answer.

_____

What is the probability you will pick a white shirt from the box and then a pick blue shirt from the box without looking? _____

Answer 8: Event A and Event B are dependent events because the shirt chosen in Event A is NOT put back into the box. Therefore, there are fewer shirts in the box during Event B.

The probability you will pick a white shirt from a box and then pick a blue shirt from a box without looking is 1/15.

Explanation: P(white shirt) = 6/10 = 3/5. Since you don't put the white shirt back in the box there are only 9 shirts left in the box (3 blue, 1 red, and 5 white).
Therefore, P(blue shirt) = 3/9 = 1/3.

Use conditional probability to find 3/5 * 1/3 = 1/5.

****

Question 9: In how many ways can 6 books be arranged on a shelf if there is room for only 2 books? Explain two methods you can use to find the answer.

Answer 9: 30 ways

Method 1: List all the possible permutations. (Books labeled as 1, 2, 3, 4, 5, 6)

1, 2	2, 1	3, 1	4, 1	5, 1	6, 1
1, 3	2, 3	3, 2	4, 2	5, 2	6, 2
1, 4	2, 4	3, 4	4, 3	5, 3	6, 3
1, 5	2, 5	3, 5	4, 5	5, 4	6, 4
1, 6	2, 6	3, 6	4, 6	5, 6	6, 5

Method 2: Use the Permutation Formula: $_n P_r = \dfrac{n!}{(n-r)!}$ where n = 6 and r = 2.

$$_5 P_2 = \frac{6!}{(6-2)!} = \frac{6!}{4!} = \frac{6*5*4*3*2*1}{4*3*2*1} = 6*5 = 30 \text{ ways}$$

****

Question 10: How many combinations can be made from 6 books taken 2 books at a time? Explain two methods you can use to find the answer.

Answer 10: 15 combinations
Method 1: List all the possible combinations.

1, 2	2, 3	3, 4	4, 5	5, 6
1, 3	2, 4	3, 5	4, 6	
1, 4	2, 5	3, 6		
1, 5	2, 6			
1, 6				

Method 2: Use the Combination Formula: $\begin{bmatrix} n \\ k \end{bmatrix} = \dfrac{n!}{k!(n-k)!}$ where n = number of distinct items and k = number of items in each group.

$$\begin{bmatrix} 6 \\ 2 \end{bmatrix} = \frac{6!}{2!(6-2)!} = \frac{6!}{2!(4)!} = \frac{6*5*4*3*2*1}{2*1*4*3*2*1} = \frac{6*5}{2*1} = \frac{30}{2} = 15 \text{ combinations}$$

****

Question 11: Explain why the Answer to Question 9 and the Answer to Question 10 were different.

Answer 11: Question 9 dealt with specific arrangement of the books so the order mattered. When the order matters you find the number of Permutations.

Question 10 dealt with combinations where order does not matter. When order doesn't matter you find the number of Combinations.

****

Question 12: Given the following set of data: {18, 97, 15, 30, 18, 24, 59, 61}

What is the mean of the data? _____
What is the mode of the data? _____
What is the median of the data? _____
What is the range of the data? _____

Answer 12:
What is the mean of the data? **Mean = 40.25**
What is the mode of the data? **Mode = 18**
What is the median of the data? **Median = 27**
What is the range of the data? **Range = 82**

****

Question 13: Given the following set of data: {1, 3, 10, 15}

What is the mean of the data? _____
What is the standard deviation of the data? _____
Explain your answers. _____

Answer 13:
What is the mean of the data? **Mean = (1 + 3 + 10 + 15)/4 = 29/4 = 7.25**

What is the standard deviation of the data? Use the formula for standard deviation:

$$\sqrt{\frac{\sum_{i=1}^{n}(X_i - \bar{X})^2}{N}}$$

Where $\bar{X}$ = the Arithmetic Mean of the data set

x_i = the value of each number in the data set

N = Number of values in the data set

$$s = \sqrt{\frac{(1-7.25)^2 + (3-7.25)^2 + (10-7.25)^2 + (15-7.25)^2}{4}} = \sqrt{\frac{124.75}{4}} = \sqrt{31.1875} = 5.58$$

****

Question 14: Keisha earned the following test scores on her last four science tests: 70, 85, 72, and 95. Keisha wants to earn an average of 80 or above in science. If there is one science test left, what score must Keisha earn in order to ensure she earns and average of 80 in science? Explain your answer.

Answer 14: She must earn a 78 or above on the last science test.
Explanation: Let "x" represent the score Keisha needs to earn on the last science class.

The mean is calculated by: $\dfrac{70+85+72+95+x}{5} = 80$

Solve for x.

$322 + x = 400$

$x = 78$

****

Question 15: Find the mode and median for each set of numbers.

{5, 18, 31} _____
{9, 9, 24, 31} _____
{2, 8, 27, 8, 2, 39, 8, 41} _____
{10, 10, 10, 10} _____

Answer 15:
{5, 18, 31} **Mode = None; Median = 18**
{9, 9, 24, 31} **Mode = 9; Median = 16.5**
{2, 8, 27, 8, 2, 39, 8, 41} **Mode = 8; Median = 8**
{10, 10, 10, 10} **Mode = 10; Median = 10**

Additional Topics From Algebra And Geometry

What are Complex numbers? Complex numbers involve imaginary numbers. Imaginary numbers are represented by i. An imaginary number would be used when you are trying to find square root of a negative number. (i.e. $\sqrt{-4}$).

You will need to memorize the following regarding imaginary numbers:

$i = \sqrt{-1}$ $\qquad$ $i^2 = -1$ $\qquad$ $i^3 = -i$ $\qquad$ $i^4 = 1$ $\qquad$ $i^5 = i$

A complex number is written in the form a + bi where "a" and "b" are real numbers and $i = \sqrt{-1}$. "a" is referred to as the real part of the number and "b" is referred to as the imaginary part of the number.

How do you work with Complex numbers? Follow these rules for complex numbers regarding each basic operation.

1. ADDITION OF COMPLEX NUMBERS:
Add the "real" parts and "imaginary" parts separately.

For example,
$(3 + 4i) + (7 + 12i) = (3+7) + (4+12)i = 10 + 16i$

2. SUBTRACTION OF COMPLEX NUMBERS:
Subtract the "real" parts and "imaginary" parts separately.

For example,
$(3 + 4i) - (7 + 12i) = (3-7) + (4-12)i = -4 - 8i$

3. MULTIPLICATION OF COMPLEX NUMBERS:
Multiply the terms by using the FOIL method and replace i^2 with a -1.

For example,
$(5 + 3i)(2 - 2i) = 10 - 10i + 6i - 6i^2 = 10 - 4i - 6(-1) = \mathbf{16 - 4i}$

4. DIVISION OF COMPLEX NUMBERS:
Multiply the numerator and denominator of the fraction by the conjugate of the denominator. Note: The conjugate of a + bi is a – bi.

For example,

$$\frac{2+i}{3-4i} = \frac{2+i}{3-4i} * \frac{3+4i}{3+4i} = \frac{2+11i}{25}$$

"TRY IT YOURSELF" – QUESTIONS ABOUT ADDITIONAL TOPICS.

Question 1: Evaluate $(10 + 2i) + (4 + 7i)$. Explain your answer.

Answer 1: $14 + 9i$
Explanation: Add the real numbers and the imaginary numbers. The imaginary numbers have an i in their structure.

$(10 + 2i) + (4 + 7i)$
(real numbers) + (imaginary numbers)
$(10 + 4) + (2i + 7i) = 14 + 9i$

****

Question 2: Evaluate $(75 + 12i) – (30 + 3i)$.

Answer 2: $45 + 9i$

****

Question 3: Evaluate $(-44 – 2i) – (-21 – 6i)$.

Answer: $-23 + 4i$
Make sure to keep track of your positive and negative signs using the rules for positive and negative signs.

****

Question 4: Evaluate $(2 + 10i)(5 + 2i)$. Explain your answer.

Answer 4: -10 + 54*i*

Explanation: Multiply the terms using the FOIL method. Replace i^2 with -1.
$(2 + 10i)(5 + 2i)$
$10 + 4i + 50i + 20i^2$
$10 + 54i + 20(-1)$
$-10 + 54i$

****

Question 5: Evaluate $\dfrac{4 - 5i}{2 + 7i}$. Explain your answer.

Answer 5: Multiply the numerator and the denominator of the fraction by the conjugate of the denominator.

$$\frac{4-5i}{2+7i} = \frac{4-5i}{2+7i} * \frac{2-7i}{2-7i} = \frac{8-28i-10i+35i^2}{4-14i+14i-49i^2} = \frac{8-38i-35}{4+49} = \frac{-27-38i}{53}$$

****

Question 6: Write $3^2 = 9$ using logarithmic notation.

Answer 6: 2 is the logarithm of 9 to the base 3 therefore logarithmic notation is

$2 = \log_3 9$.

****

Question 7: Evaluate $\log_3 81$. Explain your answer.

Answer 7: $\log_3 81$ says that you have a base = 3 and you have to figure out what x value to use in order to satisfy $3^x = 81$. Therefore, x = 4 and $\log_3 81 = 4$.

****

Question 8: If you borrow $100 at 15% interest for 1 year, how much interest will you end up paying on the loan? Explain your answer.

Answer 8: I = $15
Use the formula for Simple Interest which is I = Prt.
I = ($100)(0.15)(1) = $15

****

Question 9: What is the perimeter and area of a square with a side length of 12cm?

Answer 9: Perimeter = 48cm; Area = 144 cm2

****

Question 10: What is the perimeter of a rectangle with a length of 20ft and a width of 10ft?

Answer 10: Perimeter = 60ft; Area = 200ft

****

Question 11: What is the perimeter and area of a right triangle with side lengths 6in, 8in, and a hypotenuse of 10in?

Answer 11: Perimeter = 24in; Area = 24in2

****

Question 12: Solve $x + 22 = 99$. Explain your answer.

Answer 12: $x = 77$.
Subtract 22 from both sides of the equation to get "x" by itself.

****

Question 13: Solve $4x - 25 = 75$. Explain your answer.

Answer 13: $x = 25$.
Add 25 to both sides of the equation. Divide each side by 4.

****

Question 14: Solve $\frac{1}{4}x - 3 = 13$. Explain your answer.

Answer 14: $x = 64$
Add 3 to each side of the equation. Multiply each side of the equation by 4.

****

Question 15: Solve and graph $x - 4 > 3$.

Answer 15: $x > 7$

Note: The open circle denotes the fact that 7 is NOT included in the solutions to the inequality.

****

Question 16: Solve and graph $3x \leq 12$.

Answer 16: $x \leq 4$

Note: The circle is shaded in to denote the fact that 4 is included in the solutions to the inequality.

🎓 Logarithms and Exponents

What are logarithms? Logarithms represent the exponent of a positive number. For example, if $b^x = N$ where "N" is a positive number and "b" is a positive number besides 1, then the exponent "x" is the logarithm of N to the base b.

This relationship can be written as $x = \log_b N$

This would be said "x equals log base "b" of "N.""

Following are examples that will help demonstrate this relationship:

Example 1: Write $4^2 = 16$ using logarithmic notation.

2 is the logarithm of 16 to the base 4 therefore logarithmic notation is $2 = \log_4 16$

Example 2: Evaluate $\log_4 64$.

$\log_4 64$ says that you have a base $= 4$ and you have to figure out what x value to use in order to satisfy $4^x = 64$. Therefore, x $=3$ and $\log_4 64 = 3$.

LAWS OF LOGARITHMS: There are 3 basic laws of logarithms which include:

1. The logarithm of the product of two positive numbers M and N is equal to the sum of the logarithms of the numbers

$$\log_b MN = \log_b M + \log_b N$$

For example, $\log_2 3(5) = \log_2 3 + \log_2 5$

2. The logarithm of the quotient of two positive numbers M and N is equal to the difference of the logarithms of the numbers

$$\log_b \frac{M}{N} = \log_b M - \log_b N$$

For example, $\log_{10} \frac{17}{24} = \log_{10} 17 - \log_{10} 24$

3. The logarithm of the "pth" power of a positive number M is equal to "p" multiplied by the logarithm of the number

$$\log_b M^p = p \log_b M$$

For example, $\log_8 6^4 = 4 \log_8 6$

What are Natural Logarithms? Natural logarithms are logs with a base "e" which is a constant. Natural logarithms are denoted by ln.

You can find "e" on your scientific calculator. e $= 2.718281828\ldots$

$y = 15x$ is $x = y^2$

When would you use logarithms and exponents? Logarithms and exponents can be used when you calculate simple & compound interest and exponential growth.

Example of Simple Interest:
The formula for Simple Interest is I = Prt where I = simple interest, P = principal, r = annual interest rate, and t= time.

If you borrow $400 at 10% interest for 1 year, how much interest will you end up paying on the loan?

I = Prt
I = ($400)(0.10)(1)
I = $40 in interest

Example of Compound Interest:
Compound interest is paid periodically over the term of a loan. This gives a new principal amount at the end of <u>each</u> interval of time.

Use $A = Pe^{rt}$ when interest in compounded continuously where A = Total amount owed (principal plus interest), P = Principal, r = annual interest rate, t = years

Find the amount of an investment if $10,000 is invested at 8% compounded continuously for 2 years.

$$A = Pe^{rt}$$

$$A = 10,000e^{(0.08)(2)}$$

$$A = \$11,730$$

 # Perimeter and Area of Plane Figures

How do you find the Perimeter of a figure? The Perimeter is the SUM of the length of each side of the figure.

Following are the Perimeter formulas for a Square, Rectangle, Triangle, and Circle.

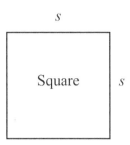

Perimeter of a Square = s + s + s + s = 4s

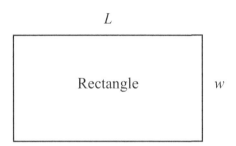

Perimeter of a Rectangle = L + w + L + w = 2L + 2w

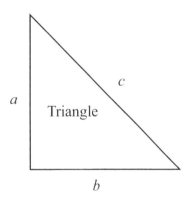

Perimeter of a Triangle = a + b + c

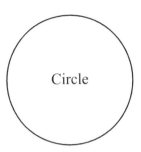

Note: The Perimeter of a Circle is referred to as the "Circumference"

Circumference = $2\pi r$

Where r = radius of the circle

How do you find the Area of a figure? Area represents the entire region within the perimeter of the figure.

Following are Area formulas for a Square, Rectangle, and Triangle. The colored region represents the "Area" of the figure.

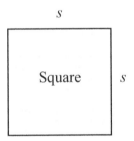

Area of a Square = s * s = s^2

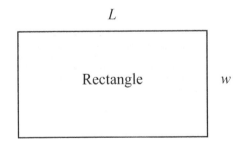

Area of a Rectangle = L * w

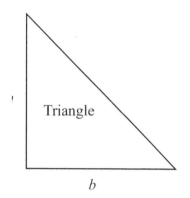

Area of a Triangle = ½ bh

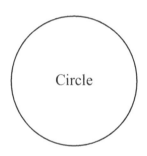

Area of a Circle = πr²

🎓 The Pythagorean Theorem

What is the Pythagorean Theorem? The Pythagorean Theorem is used when you have a right triangle. This theorem allows you to determine the length of one side based on the lengths of the other two sides of the triangle.

The Pythagorean Theorem states that the length of the hypotenuse squared is equal to the sum of the squares of the lengths of the legs: (a & b are the legs. c is the hypotenuse.)

$$a^2 + b^2 = c^2$$

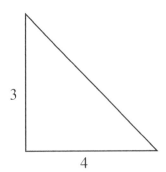

Example of the Pythagorean Theorem: The legs of a triangle are 3 inches and 4 inches respectively. Find the length of the hypotenuse.

We know the legs are a = 3 inches and b = 4 inches, so solve for c.

$$a^2 + b^2 = c^2$$

$$3^2 + 4^2 = c^2$$

$$9 = 16 = c^2$$

$$25 = c^2$$

Therefore, **c = 5 inches**

🎓 *Parallel and Perpendicular Lines*

What are parallel lines? Two lines are parallel if they lie in the same plane and never intersect. Parallel lines are denoted by the $\|$ symbol.

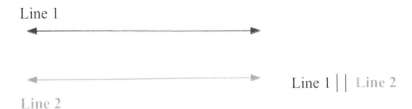

Line 1

Line 2

Line 1 $\|$ Line 2

What are perpendicular lines? Two lines are perpendicular if their intersection forms a right angle. Perpendicular lines are denoted by the ⊥.

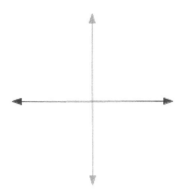

Line 1 ⊥ Line 2

🎓 Algebraic Equations

What is an algebraic equation? Equations contain numbers and variables that are called "terms." An algebraic equation must have an equal sign. Equations must always be in "balance" meaning that the value of each side of the equation is the same.

How do you solve an equation? You need to follow specific procedures when solving an equation. The two main goals you need to accomplish include moving the variable to one side of the equal sign and setting the coefficient of the variable equal to 1.

Example of a "One" Step Equation: Solve x + 10 = 28

The variable "x" already has a coefficient of 1 so we need to move 10 to the other side of the equation. Therefore, you will add the opposite of 10 to each side of the equation. Remember: When you complete an operation on one side of the equation you MUST do the exact same operation to the other side of the equation.

$$x + 10 = 28$$
$$x + (10 - 10) = (28 - 10)$$
$$x + 0 = 18$$
$$x = 18$$

Example of a "Two" Step Equation: Solve 2x + 15 = 45

First, move 15 to the other side of the equation so that 2x is by itself on one side.

$$2x + 15 = 45$$
$$2x + (15 - 15) = (45 - 15)$$
$$2x = 30$$

The second step is to make the coefficient of the variable 1. Therefore, divide each side by 2.

$$2x = 30$$

$$\frac{2x}{2} = \frac{30}{2} = 15$$

Therefore, $x = 15$

Note: Dividing by 2 is the equivalent to multiplying its reciprocal of ½.

Example of "Two" Step Equation using a Reciprocal: Solve $\frac{2}{3}x + 2 = 8$

$$\frac{2}{3}x + 2 = 8$$

$$\frac{2}{3}x + (2 - 2) = (8 - 2)$$

$$\frac{2}{3}x = 6$$

$$\frac{3}{2} * \frac{2}{3}x = 6 * \frac{3}{2}$$

$$x = 9$$

 # Algebraic Inequalities

What are the four equality signs? Read the following signs from left to right.

$<$ is "less than"

$>$ is "greater than"

$\leq$ is "less than or equal to"

$\geq$ is "greater than or equal to"

You can use inequality signs in an algebraic equation format.

For example, Solve and graph x < 4 on a number line.

In equalities you will have more than one possible value for the variable x. This inequality is "solved" because x is on one side by itself and has a coefficient of 1. Therefore, you can go right to graphing it on a number line.

In this example, every number less than and NOT including 4 is a possible value. Since 4 is NOT a possible value for x then you will leave an OPEN circle at 4. Shade in the values less than 4 to represent all the possible values for x. Remember to shade in the arrow as well indicating that the possible values go on into negative infinity.

$$x < 4$$

Systems of Equations

What is a system of equations? A system of equations is just another way of saying a group of equations. Although a system of equations can include more than two equations, the term generally refers to just two. An equation with two variables (such as $x + y = 1$) has infinitely many solutions, but when studying systems of equations there is only one set of coordinates that solves both equations. In other words, it is the point at which the graphs of the two equations intersect. This is called a simultaneous solution, because it applies to both equations.

How do you find a simultaneous solution? Finding the simultaneous solution is called solving the system. You do this by solving one equation and substituting it into the other.

Example: Find the simultaneous solution for $\begin{matrix} x + y = 5 \\ x - y = 1 \end{matrix}$

We must first solve for one variable in one equation:

$$x - y = 1$$
$$x = (1 + y)$$

We can then substitute it into the other equation:

$$x + y = 5$$
$$(1 + y) + y = 5$$
$$1 + 2y = 5$$
$$2y = 4$$
$$y = 2$$

We now know that $y = 2$. We can use this answer to solve for x. As you can see, it doesn't matter which equation we plug $y = 2$ into. Because the simultaneous solution holds true for both equations they will have the same answer.

$$x + y = 5$$
$$x + (2) = 5$$
$$x = 3$$
$$OR$$
$$x - y = 1$$
$$x - (2) = 1$$
$$x = 3$$

Therefore, the simultaneous solution for the system is the point $(3, 2)$. This means that the graphs of the two functions intersect at this point.

Factor Theorem

What is the Factor Theorem? The Factor Theorem states:

If ($P(a) = 0$), then ($x - a$) is a factor of ($P(x)$).
If ($x - a$) is a factor of ($P(x)$), then ($P(a) = 0$).

In other words, if the graph of a polynomial crosses the x axis (or equals zero) at a, then a is a factor of the polynomial.

How can you use the Factor Theorem? The Factor Theorem can be used determine if something is a factor of an equation.

Example: Determine whether $x+3$ is a factor of $P(x) = x^2 + 7x + 12$. According to the Factor Theorem $x+3$ is a factor of $P(x)$ if $P(-3) = 0$.

$$P(x) = x^2 + 7x + 12$$
$$P(-3) = (-3)^2 + 7(-3) + 12$$
$$P(-3) = 9 - 21 + 12$$
$$P(-3) = 0$$

Because $P(-3) = 0$, $x+3$ is a factor of $P(x) = x^2 + 7x + 12$

Fundamental Theorem of Algebra

What is the Fundamental Theorem of Algebra? The Fundamental Theorem of Algebra was proven by Carl Friedrich Gauss, and states that any polynomial $P(x)$ with a positive degree has at least one zero, or root. Zeros are any place where the graph crosses the x axis, so basically they are x intercepts.

How do you use the Fundamental Theory of Algebra? Because the Fundamental Theorem of Algebra states that every polynomial equation has zeros, we can solve for them by substituting 0 in for $f(x)$.

Example 1:

$$f(x) = x^2 - 1$$
$$0 = x^2 - 1$$
$$1 = x^2$$
$$x = 1$$

An answer of $x = 1$ means that the graph of $f(x)$ crosses the x axis where $x = 1$.

Example 2:

$$f(x) = x^2 + 9x + 20$$
$$0 = x^2 + 9x + 20$$
$$0 = (x+5)(x+4)$$
$$x+5 = 0$$
$$x = -5$$
$$x+4 = 0$$
$$x = -4$$

Therefore, the graph of $f(x)$ crosses the x axis where $x = -4$ and where $x = -5$.

Remainder Theorem

What is the Remainder Theorem? The Remainder Theorem states that If $P(x)$ is a polynomial and a is any number, then $\dfrac{P(x)}{x-a} = P(a)$.

How do you use the Remainder Theorem? The Remainder Theorem can be used to determine the remainder of a polynomial.

Example 1: Determine the remainder that will occur when $P(x) = x^4 + 3x^3 + 4x^2 - 15x - 26$ is divided by $x - 2$.

According to the Remainder Theorem $P(2)$ should calculate the remainder:

$$P(x) = x^4 + 3x^3 + 4x^2 - 15x - 26$$
$$P(2) = (2)^4 + 3(2)^3 + 4(2)^2 - 15(2) - 26$$
$$P(2) = 16 + 3(8) + 4(4) - 30 - 26$$
$$P(2) = 16 + 24 + 16 - 30 - 26$$
$$P(2) = 0$$

Therefore the remainder is 0.

Example 2: Determine the remainder that will occur when
$P(x) = x^4 + 3x^3 + 4x^2 - 15x - 26$ is divided by $x - 3$.

This time we will substitute 3 into the equation:

$P(x) = x^4 + 3x^3 + 4x^2 - 15x - 26$

$P(3) = (3)^4 + 3(3)^3 + 4(3)^2 - 15(3) - 26$

$P(3) = 81 + 3(27) + 4(9) - 45 - 26$

$P(3) = 81 + 81 + 36 - 45 - 26$

$P(3) = 127$

Therefore the remainder is 127.

Inequality Example: Solve and graph x + 10 < 5
Solve this as you would an equation keeping the < sign in tact.

$$x + 10 < 5$$
$$x + (10 - 10) < (5 - 10)$$
$$x < -5$$

Inequality Example: Solve and graph 2x + 4 ≥ 10

$$2x + 4 \geq 10$$
$$2x + (4 - 4) \geq (10 - 4)$$
$$2x \geq 6$$
$$\frac{2x}{2} \geq \frac{6}{2}$$
$$x \geq 3$$

Note: Since the inequality sign is "greater than or equal to" we MUST include 3 on our number line. Therefore, you need to shade in the circle at x=3.

Special Rule when solving Inequalities: There is one special rule that you need to use when you have a –x value to begin with. Remember that we want x to be positive and have a coefficient of 1. In the very last step of the inequality you divide each side of the inequality by a negative number. This means that you need to change the direction of the inequality sign to the opposite direction. This is ONLY when dividing (or multiplying by a reciprocal) by a NEGATIVE number.

Special Rule Inequality Example: Solve and graph -2x < 10.

$$-2x < 10$$
$$\frac{-2x}{-2} < \frac{10}{-2}$$
$$x > -5$$

Sample Questions

Section 1: Sets

1. If $A = \{2,4,6,8\}$ and $B = \{1,3,5,7\}$, then what is $A \bigcup B$?

   A. {1, 2, 3, 4, 5, 6, 7, 8}
   B. {}
   C. {1, 2, 3, 4}
   D. {1, 2, 5, 6}
   E. {1, 2, 3, 4, 5, 6, 7}

2. If $A = \{2,4,6,8\}$ and $B = \{1,3,5,7\}$, then what is $A \bigcap B$?

   A. {1, 2, 3, 4, 5, 6, 7, 8}
   B. {}
   C. {1, 2, 3, 4}
   D. {1, 2, 5, 6}
   E. {1, 2, 3, 4, 5, 6, 7}

3. Which of the following is an infinite set?

   A. {1, 2, 3, 4, 5, 6}
   B. {2, 4, 6, 8, …}
   C. {1, 2, 3, 5, 6, 7}
   D. {1, 14, 18, 20}
   E. None of the above are infinite sets

4. Which of the following is a finite set?

   A. {1, 2, 3, 4, 5, …}
   B. {2, 22, 222, 2222, …}
   C. {1, 3, 5, 7, 9, …}
   D. {2, 4, 6, 8}
   E. None of he above are finite sets.

5. If $A = \{2,4,6,8\}$ and $B = \{1,3,5,7\}$, then $A \bigcap B$ is a(n)…?

   A. Infinite set
   B. Equivalent set
   C. Empty set
   D. Finite set
   E. Function

6. Which of the following sets are disjoint sets?

   A. A={1, 2, 3, 4}, B={1, 2, 3, 4}
   B. A={1, 2, 3, 4}, B={2, 3, 4}
   C. A={2, 4, 6}, B={1, 3, 5, 7}
   D. A={2, 3, 4, 9}, B={25, 67, 34, 9}
   E. All of the above are disjoint sets

7. What is $A \times B$ if A={1, 2, 3} and B={1, 2, 3}?

   A. {1, 2, 3}
   B. {6}
   C. {(1,0) ,(1,2), (1,3), (2,0), (2,2), (2,3), (3,0), (3,2), (3,3)}
   D. {(1,1) ,(1,2), (1,3), (2,1), (2,2), (2,3), (3,1), (3,2), (3,3)}
   E. {36}

8. What is represented by the overlapping area in a Venn Diagram?

   A. The common elements between the sets.
   B. The elements which belong only to the first set.
   C. The elements which belong to neither set.
   D. The elements which belong only to the second set.
   E. There is no overlapping area in Venn Diagrams.

9. {(1,0) ,(1,7), (1,3), (5,0), (5,7), (5,3), (3,0), (3,7), (3,3)} is the Cartesian product of ___?

   A. It is not a Cartesian product, it is an intersection.
   B. {0 1, 3, 5, 7}
   C. {1, 3, 3} and {0, 5, 7}
   D. {0, 1, 2} and {3, 4, 5}
   E. {1, 5, 3} and {0, 7, 3}

10. Which of the following are equivalent sets?

    A. A={1, 2, 3, 4}, B={1, 2, 3, 4}
    B. A={1, 2, 3, 4}, B={2, 3, 4}
    C. A={2, 4, 6}, B={1, 3, 5, 7}
    D. All of the above are disjoint sets
    E. All of the above are equivalent sets

Section 2: Logic

For problems 1-4 use the following statement:

All boys are taller than Sally.

1. Which of the following the converse of the statement?

    A. Tom is a boy, and he is shorter than Sally.
    B. Sally is shorter than all boys.
    C. A person who is not a boy is not taller than Sally.
    D. All boys are not taller than Sally.
    E. None of the above

2. Which of the following is the Negative of the statement?

    A. Tom is a boy, and he is shorter than Sally.
    B. Sally is shorter than all boys.
    C. A person who is not a boy is not taller than Sally.
    D. All boys are not taller than Sally.
    E. None of the above

3. Which of the following is a counterexample of the statement?

    A. Tom is a boy, and he is shorter than Sally.
    B. Sally is shorter than all boys.
    C. A person who is not a boy is not taller than Sally.
    D. All boys are not taller than Sally.
    E. None of the above

4. Which of the following is the Inverse of the statement?

    A. Tom is a boy, and he is shorter than Sally.
    B. Sally is shorter than all boys.
    C. A person who is not a boy is not taller than Sally.
    D. All boys are not taller than Sally.
    E. None of the above

5. Which of the following connectives is a conjunction?

    A. and
    B. or
    C. not
    D. …if and only if…
    E. If…then …

6. Which of the following connectives is a conditional?

   A. and
   B. or
   C. not
   D. …if and only if…
   E. If…then …

7. Which of the following connectives is a disjunction?

   A. and
   B. or
   C. not
   D. …if and only if…
   E. If…then …

8. Which of the following connectives is a negation?

   A. and
   B. or
   C. not
   D. …if and only if…
   E. If…then …

9. Which of the following would prove a hypothesis necessary and sufficient for its conclusion?

   A. If a statement and its converse are both true.
   B. If a statement and its converse are both false.
   C. If a statement is true, and its converse is false.
   D. If a statement is false, and its converse is true.
   E. None of the above

10. Which of the following would prove a hypothesis necessary but not sufficient for its conclusion?

    A. If a statement and its converse are both true.
    B. If a statement and its converse are both false.
    C. If a statement is true, and its converse is false.
    D. If a statement is false, and its converse is true.
    E. None of the above

Section 3: Real Number System

1. Which of the following is NOT a prime number?

    A. 1
    B. 2
    C. 3
    D. 5
    E. 7

2. What is a composite number?

    A. A number that has more than one digit.
    B. A number that has only one digit.
    C. A number which is not divisible by two.
    D. Any number which is not prime.
    E. Another name for a prime number.

3. Which of the following numbers is even?

    A. 11
    B. 2
    C. 3
    D. 5
    E. All of the above are odd numbers

4. What is the Greatest Common Factor of 48 and 36?

    A. 2
    B. 3
    C. 4
    D. 6
    E. 12

5. Which of the following correctly lists all of the factors of 36?

    A. 1, 36
    B. 1, 2, 18, 36
    C. 1, 2, 3, 5, 12, 18, 36
    D. 1, 2, 3, 4, 6, 9, 12, 18, 36
    E. 1, 2, 3, 4, 5, 6, 9, 12, 18, 36

6. Which of the following is NOT a divisor of 15?

    A. 45
    B. 15
    C. 5
    D. 3
    E. 1

7. $|-5| = ?$

    A. -5
    B. 5
    C. 1
    D. -1
    E. None of the above

8. $-|-7| = ?$

    A. 49
    B. 7
    C. -7
    D. -1(-7)
    E. None of the above

9. $|3 - 13| = ?$

    A. 16
    B. -16
    C. 10
    D. -10
    E. 39

10. Which of the following is true of closed intervals?

    A. Closed intervals do not include the stated endpoints.
    B. Closed intervals are more common than open intervals.
    C. Closed intervals are less common than open intervals.
    D. When graphed, a closed interval is signified by a circle which is not shaded.
    E. When graphed, a closed interval is signified by a shaded circle.

11. Which of the following numbers is irrational?

    A. .9
    B. .134526
    C. π
    D. 0
    E. 1

12. To which of the following groups does the number 5 NOT apply?

    A. Integers
    B. Irrational Numbers
    C. Whole Numbers
    D. Natural Numbers
    E. Rational Numbers

13. To which of the following groups does 0 not apply?

    A. Whole Numbers
    B. Integers
    C. Rational Numbers
    D. Natural Numbers
    E. 0 applies to all of the above

14. Which of the following is NOT a divisor of 48?

    A. 5
    B. 4
    C. 6
    D. 8
    E. 12

15. Determine which numbers in the following set are irrational: $\{1, 2, \sqrt{4}, \sqrt{5}, \frac{3}{4}\}$

    A. 1
    B. 2
    C. $\sqrt{4}$
    D. $\sqrt{5}$

    E. $\frac{3}{4}$

16. Which of the following numbers is even?

    A. 1
    B. 3
    C. 47
    D. 69
    E. All of the above are odd numbers.

17. Which of the following correctly lists all of the factors of 20?

    A. 1, 20
    B. 1, 2, 10, 20
    C. 1, 2, 4, 5, 10, 20
    D. 1, 2, 3, 4, 5, 10, 20
    E. 1, 2, 3, 4, 5, 10, 12, 20

18. $\left| 3^2 + 4 - 20 \right| = ?$

    A. -7
    B. -33
    C. 20
    D. 33
    E. 7

19. Which of the following types of numbers is not included in the real number system?

    A. Natural Numbers
    B. Complex Numbers
    C. Integers
    D. Irrational Numbers
    E. Whole Numbers

20. Which of the following is a composite number?

    A. 4
    B. 3
    C. 2
    D. 1
    E. All of the above are composite

Section 4: Functions and their graphs

1. Solve, $f(x) = 13x - 5$ for $x = 3$.

    A. 8
    B. 21
    C. 34
    D. 47
    E. 65

2. $f(3) = x^2 + 4x + 3$, $x = ?$

    A. 0
    B. 24
    C. 18
    D. 6
    E. -6

3. What is the Vertical Line Test used to determine?

    A. Whether or not the graph has a defined slope.
    B. If a graph is a vertical line.
    C. If a graph is a horizontal line.
    D. Whether or not a graph represents a function.
    E. There is no such thing as the Vertical Line Test.

4. Solve $f(x) = 3x^2 + 2x + 3$ for $x = 1$

    A. 7
    B. 8
    C. 9
    D. 10
    E. 11

5. $f(5) = |3x - 14| = ?$

    A. -1
    B. 1
    C. 0
    D. 2
    E. -2

6. $f(2) = 3x - 1 = ?$

    A. 3
    B. 4
    C. 5
    D. 6
    E. 7

7. What is the domain of a function?

    A. The x values that can be input into an equation.
    B. The x values that cannot be put into an equation.
    C. The type of graph that the function produces.
    D. The y values that cannot be put into an equation.
    E. The y values that can be input into an equation.

8. What is the range of a function?

    A. The x values that can be input into an equation.
    B. The x values that cannot be put into an equation.
    C. The type of graph that the function produces.
    D. The y values that cannot be put into an equation.
    E. The y values that can be input into an equation.

9. Which of the following is NOT true?

    A. A function can have many output values for each value input.
    B. To find the square of a number you must multiply it by itself.
    C. Range is a measure of the y values an equation produces.
    D. Domain is a measure of the x values an equation allows.
    E. All of the above statements are true.

For questions 10-13 use $f(x) = 3x - 1$ and $g(x) = 4x$

10. Determine $f(g(x))$

    A. $12x - 1$
    B. $7x - 1$
    C. $12x - 4$
    D. $-x - 1$
    E. $-7x - 1$

11. Determine $g(f(x))$

    A. $12x-1$
    B. $7x-1$
    C. $12x-4$
    D. $-x-1$
    E. $-7x-1$

12. Determine $f(x)+g(x)$

    A. $12x-1$
    B. $7x-1$
    C. $12x-4$
    D. $-x-1$
    E. $-7x-1$

13. Determine $f(x)-g(x)$

    A. $12x-1$
    B. $7x-1$
    C. $12x-4$
    D. $-x-1$
    E. $-7x-1$

14. $f(x)=x^2$ is symmetric about the ____?

    A. x axis
    B. origin
    C. line x=1
    D. line y=1
    E. y axis

15. What is the inverse of $f(x)=x^2+1$?

    A. $\sqrt{x-1}$
    B. $x+2$
    C. $\sqrt{x+1}$
    D. $x-2$
    E. $\sqrt{x+2}$

16. What is the inverse of $f(x) = x^2 - 1$?

    A. $\sqrt{x-1}$
    B. $x+2$
    C. $\sqrt{x+1}$
    D. $x-2$
    E. $\sqrt{x+2}$

17. What is the inverse of $f(x) = x + 2$?

    A. $\sqrt{x-1}$
    B. $x+2$
    C. $\sqrt{x+1}$
    D. $x-2$
    E. $\sqrt{x+2}$

18. What is the inverse of $f(x) = x - 2$?

    A. $\sqrt{x-1}$
    B. $x+2$
    C. $\sqrt{x+1}$
    D. $x-2$
    E. $\sqrt{x+2}$

19. What is demonstrated by the graph below?

    A. A horizontal translation
    B. A vertical translation
    C. Symmetry about the y axis
    D. Symmetry about the x axis
    E. Symmetry about the origin

20. What is demonstrated by the graph below?

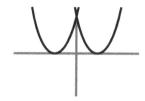

    A. A horizontal translation
    B. A vertical translation
    C. Symmetry about the y axis
    D. Symmetry about the x axis
    E. Symmetry about the origin

Section 5: Probability and Statistics

For problems 1-4 use the following data set:

$\{1, 4, 6, 8, 6, 8, 8, 47, 9\}$

1. What is the mean?

    A. 9.7
    B. 10.3
    C. 10.8
    D. 11
    E. 11.3

2. What is the median?

    A. 1
    B. 3
    C. 6
    D. 8
    E. 9

3. What is the mode?

    A. 3
    B. 6
    C. 8
    D. 9
    E. 47

4. What is the range?

    A. 8
    B. 9
    C. 47
    D. 48
    E. 46

5. Assume that all license plates have a combination of 3 numbers (0-9) and 3 letters. If no repeats are allowed, how many possible license plates are there?

    A. 17,576,000
    B. 1,700
    C. 15,767,000
    D. 11,232,000
    E. 11,000,000

6. Assume that all license plates have a combination of 3 numbers (0-9) and 3 letters. If repeats are allowed, how many possible license plates are there?

    A. 17,576,000
    B. 1,700
    C. 15,767,000
    D. 11,232,000
    E. 11,000,000

7. What is the highest a probability can be?

    A. 0
    B. 1
    C. 2
    D. 3
    E. 4

8. When using a ten sided dice what is the probability of rolling a 3?

    A. .1
    B. .2
    C. .3
    D. .5
    E. 1

9. When using an average six sided dice what is the probability of rolling two threes in a row?

   A. 30%
   B. 17%
   C. 5%
   D. 28%
   E. 2.8%

10. When rolling two ten sided dice at once, what is the probability of rolling at least one 7?

   A. .1
   B. .2
   C. .3
   D. .4
   E. .5

11. What type of graph is depicted below?

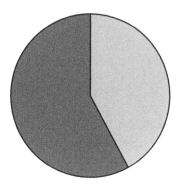

   A. Histogram
   B. Table
   C. Pie Chart
   D. Bar Graph
   E. Scatterplot

12. Which of the following statements is incorrect?

   A. The more compact a data set is the smaller the standard deviation will be.
   B. The standard deviation is a measure of how spread out the data is.
   C. Standard deviation is calculated using the mean.
   D. The more spread out the data is the smaller the standard deviation will be.
   E. All of the statements above are incorrect.

13. What is the standard deviation of the following data set?

$\{1, 5, 9\}$

    A. 1
    B. 2
    C. 3
    D. 4
    E. 5

For questions 14-16 use the following situation.

A bag is filled with 15 marbles. 5 marbles are green, 7 marbles are yellow, 2 marbles are red, and one marble is white. Marbles are picked out one at a time randomly, and replaced after each time.

14. What is the probability that a green marble will not be chosen on the first pick?

    A. 67%
    B. 5%
    C. 13%
    D. 33%
    E. 11%

15. What is the probability of picking a green marble for both of the first two draws?

    A. 67%
    B. 5%
    C. 13%
    D. 33%
    E. 11%

16. What is the probability that the sequence of the first three draws will be green, red, white?

    A. 3%
    B. 5%
    C. 13%
    D. .3%
    E. 33%

17. What type of graph is depicted below?

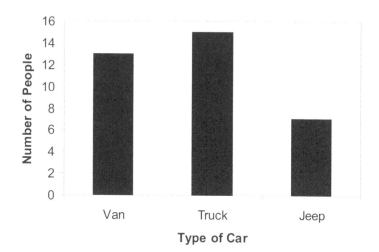

A. Histogram
B. Bar Graph
C. Box plot
D. Pie Chart
E. Line Graph

18. What is the formula for finding a permutation?

A. $_nP_r = \dfrac{n!}{(n-r)!}$

B. $_nP_r = \dfrac{n!}{n!-r!}$

C. $_rP_n = \dfrac{n!}{n!-r!}$

D. $_nP_r = \dfrac{n!}{n!-r!}$

E. $_nP_r = \dfrac{n!}{n-r}$

19. What is a permutation?

 A. An ordered sequence of items taken from a set of distinct items with replacement.
 B. A permutation is another name for a sequence called a combination.
 C. An ordered sequence of items taken from set of distinct items without replacement.
 D. All of the above statements correctly describe permutations.
 E. None of the above statements describe permutations.

20. What is the formula for a combination?

 A. $_nC_k = \dfrac{n!}{(n-k)}$

 B. $_nC_k = \dfrac{n!}{k!(n-k)}$

 C. $_kC_n = \dfrac{n!}{k!(n-k)!}$

 D. $_nC_k = \dfrac{n!}{(n-k)!}$

 E. $_nC_k = \dfrac{n!}{k!(n-k)!}$

21. What type of graph is depicted below?

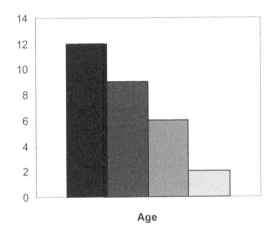

 A. Bar Graph
 B. Line Graph
 C. Pie Chart
 D. Scatterplot
 E. Histogram

For questions 22-24 use the following situation:

There are 20 random mismatched socks in a drawer. 12 of the socks are white, 3 of the socks are black, 2 of the socks are blue, and the remaining 3 are striped. Socks are randomly picked from the drawer, each having an equal chance of being chosen. Assume the socks are NOT replaced between each pick.

22. What is the probability of choosing all three striped socks in a row?

    A. .8%
    B. 10%
    C. 15%
    D. .008%
    E. 2%

23. What is the probability of choosing a white sock or a blue sock?

    A. .6
    B. .7
    C. .8
    D. .75
    E. .65

24. What is the probability of choosing first a blue or black sock, and then a white or striped sock?

    A. 17.5%
    B. 18%
    C. 18.5%
    D. 20%
    E. 70%

25. You are studying the relationship between temperature and what day of the year it is. What type of graph should you use to display your data?

    A. Pie Chart
    B. Histogram
    C. Line Graph
    D. Histogram
    E. Table

Section 6: Additional Topics From Algebra and Geometry

1. Simplify $\sqrt{-25}$

    A. $-5i$
    B. -5
    C. 5
    D. $5i$
    E. Not possible

2. Solve $\log x = 2$

    A. 100
    B. 10
    C. 1
    D. 2
    E. 0

3. Simplify $(3i+4)-(4i+2)$

    A. $i+2$
    B. $7i+6$
    C. $-i+2$
    D. $7i-6$
    E. $7i+2$

For problems 4-9 use the following figures

  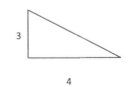

4. What is the area of the circle above?

    A. 10
    B. 12.6
    C. 12.9
    D. 13
    E. 13.5

5. What is the circumference of the circle above?

    A. 10
    B. 12.6
    C. 12.9
    D. 13
    E. 13.5

6. What is the perimeter of the rectangle above?

    A. 18
    B. 19
    C. 20
    D. 21
    E. 22

7. What is the area of the rectangle above?

    A. 18
    B. 19
    C. 20
    D. 21
    E. 22

8. Assuming that it is a right triangle, what is the value of the remaining side of the triangle above?

    A. 3
    B. 4
    C. 5
    D. 6
    E. 7

9. What is the area of the triangle above?

    A. 5
    B. 6
    C. 7
    D. 8
    E. 9

10. If a triangle has sides such that the legs are 7 and 5, then what is the value of the hypotenuse?

    A. 10
    B. 9
    C. 9.6
    D. 8
    E. 8.6

11. If a triangle has sides such that the hypotenuse is 13, and one of the legs is 12, then what is the value of the remaining leg?

    A. 8
    B. 7
    C. 6
    D. 5
    E. 4

12. What are parallel lines?

    A. Lines which lie in the same plane and never intersect.
    B. Lines which lie in different planes, and therefore never intersect.
    C. Lines for which the intersection forms a right angle.
    D. Lines for which the intersection is anything but a right angle.
    E. Parallel lines do not exist and therefore cannot be properly described.

13. What are perpendicular lines?

    A. Lines which lie in the same plane and never intersect.
    B. Lines which lie in different planes, and therefore never intersect.
    C. Lines for which the intersection forms a right angle.
    D. Lines for which the intersection is anything but a right angle.
    E. Parallel lines do not exist and therefore cannot be properly described.

14. Find the simultaneous solution for $\begin{array}{l} y = 4x + 7 \\ y = 2x + 4 \end{array}$

    A. $(-\dfrac{2}{3}, 1)$

    B. $(-\dfrac{3}{2}, 1)$

    C. $(-\dfrac{2}{3}, 2)$

    D. $(-2, 1)$

    E. $(\dfrac{2}{3}, 1)$

15. At what y coordinate do the following equations intersect?

$y = 3x$
$y = 6x - 2$

    A. 6
    B. 5
    C. 4
    D. 3
    E. 2

16. At what x coordinate do the following equations intersect?

$y = x - 1$
$y = 2x + 1$

    A. -2
    B. -1
    C. 0
    D. 1
    E. 2

17. According to the Remainder Theorem $\dfrac{P(x)}{x-a} =$

    A. $P(a)$

    B. $\dfrac{P(a)}{a}$

    C. $\dfrac{P(a)}{x}$

    D. x

    E. Not enough information

For problems 18-19 use the following equation:

$$P(x) = x^4 - 2x^3 + 3x^2 - 2x$$

18. Find the remainder of $\dfrac{P(x)}{x-1}$

    A. 4

    B. 3

    C. 2

    D. 1

    E. 0

19. Is $x-1$ a factor of $P(x)$?

    A. Yes, because $P(1) = 0$

    B. Yes, because $P(-1) = 0$

    C. No, because $P(1) = 0$

    D. No, because $P(-1) = 0$

    E. No, because $P(1) = 0$

20. At what x values does $f(x) = x^2 + 3x - 4$ cross the x axis?

    A. 1 and 4

    B. -1 and 4

    C. -1 and -4

    D. 1 and -4

    E. Never

Answer Key

Section 1: Sets

1. A	5. C	9. E
2. B	6. C	10. A
3. B	7. D	
4. D	8. A	

Section 2: Logic

1. B	5. A	9. A
2. D	6. E	10. C
3. A	7. B	
4. C	8. C	

Section 3: Real Number System

1. A	6. A	11. C	16. E
2. D	7. B	12. B	17. C
3. B	8. C	13. D	18. E
4. E	9. C	14. A	19. B
5. D	10. E	15. D	20. A

Section 4: Functions and Their Graphs

1. C	6. C	11. C	16. C
2. B	7. A	12. B	17. D
3. D	8. E	13. D	18. B
4. B	9. A	14. E	19. E
5. B	10. A	15. A	20. A

Section 5: Probability and Statistics

1. C	8. A	15. E	22. D
2. D	9. E	16. D	23. B
3. C	10. B	17. B	24. D
4. E	11. C	18. A	25. C
5. D	12. D	19. C	
6. A	13. D	20. E	
7. B	14. A	21. E	

Section 6: Additional Topics from Algebra and Geometry

1. D	6. A	11. D	16. A
2. A	7. C	12. A	17. A
3. C	8. C	13. C	18. E
4. B	9. B	14. B	19. A
5. B	10. E	15. E	20. D

Test Taking Strategies

Here are some test taking strategies that are specific to this test and to other CLEP tests in general:

- Keep your eyes on the time
- Read the entire question and read all the answers. Many questions are not as hard to answer as they may seem. One example is a question I read about a scientific process I didn't know and couldn't even pronounce. I was thinking I would have to skip the question. But instead of skipping it like I usually would, I read it through and realized I didn't need to know that word or process at all. All I needed to do is read the chart.
- If you don't know the answer immediately, the new computer-based testing lets you mark questions and come back to them later.
- Read the wording carefully. Some words can give you hints to the right answer. There are no exceptions to an answer when there are words in the question such as "always" "all" or "none." If one of the answer choices includes most or some of the right answers, but not all, then that is not the answer. Here is an example:

  The primary colors include all of the following:
  - A) Red, Yellow, Blue, Green
  - B) Red, Green, Yellow
  - C) Red, Orange, Yellow
  - D) Red, Yellow, Blue
  - E) None of the above

Although item A includes all the right answers, it also includes an incorrect answer, making it incorrect. If I wasn't reading carefully, were in a hurry, or didn't know the material well, I might fall for this.

- Make a guess, there is no penalty for an incorrect answer.

 What Your Score Means

Based on your score, you may, or may not, qualify for credit at your specific institution. At University of Phoenix, a score of 50 is passing for full credit. At Utah Valley State College, the score is unpublished, the school will accept credit on a case-by-case basis. Another school, Brigham Young University (BYU) does not accept CLEP credit. To find out what score you need for credit, you need to get that information from your school's website or academic advisor.

You can score between 20 and 80 on any CLEP test. Some exams include percentile ranks. Each correct answer is worth one point. You lose no points for unanswered or incorrect questions.

Test Preparation

How much you need to study depends on your knowledge of a subject area. If you are interested in literature, took it in school, or enjoy reading then your studying and preparation for the literature or humanities test will not need to be as intensive as someone who is new to literature.

This book is much different than the regular CLEP study guides. This book actually teaches you the information that you need to know to pass the test. If you are particularly interested in an area, or you want more information, do a quick search online. We've tried not to include too much depth in areas that are not as essential on the test. Everything in this book will be on the test. It is important to understand all major theories and concepts listed in the table of contents. It is also very important to know any bolded words.

Don't worry if you do not understand or know a lot about the area. With minimal study, you can complete and pass the test.

Legal Note

FLASHCARDS

This section contains flashcards for you to use to further your understanding of the material and test yourself on important concepts, names or dates. Read the term or question then flip the page over to check the answer on the back. Keep in mind that this information may not be covered in the text of the study guide. Take your time to study the flashcards, you will need to know and understand these concepts to pass the test.

Area of a Triangle

Area of a Circle

Pythagorean Theorem

Perimeter of a Triangle

Circumference

Area of a Square

Area of a Rectangle

Perimeter of a Square

πr^2

1/2 bh

a+b+c

$a^2 + b^2 = c^2$

S*S

2πr

s+s+s+s=4s

length * width

Perimeter of a Rectangle

$\leq$

$\geq$

$\neq$

$<$

$>$

Mode

Median

Less than or equal to

l+w+l+w

Does not equal

Greater than or equal to

Greater than

Less than

The middle value

The number that appears most number of times in the sample data

Mean

{ }

Permutation

Probability

Ordered Pair

Domain

Closed Interval

Open Interval

Empty set

Sum of all values, then divided by the number of values present

The likelihood an event will occur as expressed by ratios

An ordered sequence of items taken from a set without replacement

All the possible values you can use "x" for in the function of the ordered pair

(x,y)

Does not include endpoints

Includes endpoints

Made in the USA
Coppell, TX
31 January 2022